RELIGION AND THE ENVIRONMENT

RELIGION AND THE ENVIRONMENT

Edited by:

Oliva DSouza

SOCIETY Publishing

www.societypublishing.com

Religion and the Environment

Oliva DSouza

Society Publishing

2010 Winston Park Drive,

2nd Floor

Oakville, ON L6H 5R7

Canada

www.societypublishing.com

Tel: 001-289-291-7705
001-905-616-2116
Fax: 001-289-291-7601
Email: orders@arclereducation.com

ISBN: 978-1-77407-166-3 (Hardcover)

ABOUT THE EDITOR

Dr. Oliva DSouza is a dental surgeon with post-graduate diplomas in medical administration and medical tourism. She has completed multiple certifications in Six Sigma, 5-S, public health, organizational development and people skills. Having honed her clinical skills at the Armed Forces Institute of Dental Sciences, she practiced at various private and corporate hospitals for 15 plus years, before moving on to pursuing her passion for writing, editing and academic research. She has a keen interest in research, data analysis and the application of artificial intelligence in healthcare. She freelances as a technical and academic consultant/editor for many websites, blogs and dental product companies. She has completed the women's entrepreneurship program from Indian Institute of Management, Bangalore. She serves as a director for a startup in the artificial intelligence area and also runs a business in handcrafted, natural spa products. She enjoys sports and is a voracious reader.

TABLE OF CONTENTS

GLOSSARY

A

Abasement - the action or fact of abasing or being abased; humiliation or degradation

Ailment - an illness, typically a minor one

Asceticism - severe self-discipline and avoiding of all forms of indulgence, typically for religious reasons

Attunement – to bring into harmony

Anthropologists – someone who scientifically studies humans and their customs, beliefs, and relationships

Archaeologists - a person who studies human history and prehistory through the excavation of sites and the analysis of artefacts and other physical remains

B

Bestowed – confer or present (an honour, right, or gift)

Blossoming – producing flowers or masses of flowers

Botanists – an expert in or student of the scientific study of plants

Blasphemy – the action or offence of speaking sacrilegiously about God or sacred things; profane talk

Bureaucrats - an official in a government department, in particular one perceived as being concerned with procedural correctness at the expense of people's needs

C

Controversy – prolonged public disagreement or heated discussion

Chakra System – The Chakra System is a complex network of energy channels connecting these wheels and is mapped throughout the whole body, much like a spiritual nervous system

Crucifixion – an ancient form of execution in which a person was nailed or bound to a cros

Cremation – the disposal of a dead person's body by burning it to ashes, typically after a funeral ceremon

Coherence- the quality of being logical and consistent

Ceremonies - a formal religious or public occasion, especially one celebrating a particular event, achievement, or anniversary

Conservation – prevention of wasteful use of a resource

Capitalism – an economic and political system in which a country's trade and industry are controlled by private owners for profit, rather than by the state

Communism – a theory or system of social organization in which all property is owned by the community and each person contributes and receives according to their ability and needs

Conservationists – a person who advocates or acts for the protection and preservation of the environment and wildlife

Compassion – sympathetic pity and concern for the sufferings or misfortunes of others

Crater – a large bowl-shaped cavity in the ground or on a celestial object, typically one caused by an explosion or the impact of a meteorite

Coalification - a process in which vegetable matter becomes converted into coal of increasingly higher rank with anthracite as the final product

D

Disaster – a sudden accident or a natural catastrophe that causes great damage or loss of life

Devastation – great destruction or damage

Dwelling – a house, flat, or other place of residence

Disposition- the way in which something is placed or arranged, especially in relation to other things

Demons – an evil spirit

Dynamite – a high explosive consisting of nitroglycerine mixed with an absorbent material and typically moulded into sticks

Dogmatically – If you are dogmatic, you are certain that you are right and that everyone else is wrong

Degassing – make or become free of unwanted or excess gas

Decomposition - the state or process of rotting; decay

E

Explicitly – in a clear and detailed manner, leaving no room for confusion or doubt

Eukaryotic - An organism whose cells contain a nucleus surrounded by a membrane and whose DNA is bound together by proteins (histones) into chromosomes

Ecosystem – a biological community of interacting organisms and their physical environment

Endangered – (of a species) seriously at risk of extinction

Endurance – the capacity of something to last or to withstand wear and tear

F

Feasting – eat and drink sumptuously

Formalism - excessive adherence to prescribed forms

Fatwa – a ruling on a point of Islamic law given by a recognized authority

Fundamentalists – a person who believes in the strict, literal interpretation of scripture in a religion

Folklore - the traditional beliefs, customs, and stories of a community, passed through the generations by word of mouth

Fissures - a long, narrow opening or line of breakage made by cracking or splitting, especially in rock or earth

G

Godhead - a greatly admired or influential person

Gestures – a movement of part of the body, especially a hand or the head, to express an idea or meaning

Geologists – a person who studies geology

Geothermal – relating to or produced by the internal heat of the earth

Germination – the development of a plant from a seed or spore after a period of dormancy

Guzzlers - a person who eats or drinks something greedily

H

Hereditary - conferred by or based on inheritance

Homo sapiens – the primate species to which modern humans belong; humans regarded as a species

Hymns – a religious song or poem of praise to God or a god

Homogeneous – of the same kind; alike

Humility – the quality of having a modest or low view of one's importance

Hellish – o f or like hell

I

Interdicts – an authoritative prohibition

Incarnation – a person who embodies in the flesh a deity, spirit, or quality

Inexhaustibly - incapable of being used up

Invariance - The property of remaining unchanged regardless of changes in the conditions of measurement

L

Luster - a gentle sheen or soft glow

Liturgical- relating to liturgy or public worship

Landscapes- all the visible features of an area of land, often considered in terms of their aesthetic appeal

Logging – the activity or business of felling trees and cutting and preparing the timber

Liberation - the action of setting someone free from imprisonment, slavery, or oppression

M

Myth - a traditional story, especially one concerning the early history of a people or explaining a natural or social phenomenon, and typically involving supernatural beings or events

Monks – a member of a religious community of men typically living under vows of poverty, chastity, and obedience

Metamorphosis – the process of transformation from an immature form to an adult form in two or more distinct stages

Meteorites – a piece of rock or metal that has fallen to the earth's surface from outer space as a meteor.

Mercantilism – he economic theory that trade generates wealth and is stimulated by the accumulation of profitable balances, which a government should encourage by means of protectionism

Monasteries - a building or buildings occupied by a community of monks living under religious vows

N

Netherworld - the underworld; hell

O

Oppress - keep (someone) in subjection and hardship, especially by the unjust exercise of authority

Ostensibly - as appears or is stated to be true, though not necessarily so; apparently

P

Prophet – a person regarded as an inspired teacher or proclaimer of the will of God

Precincts- the area within the walls or perceived boundaries of a particular building or place

Pilgrimage – a visit to a place that is considered special, where you go to show your respect

Plantations - an estate on which crops such as coffee, sugar, and tobacco are grown

Pollinator – A pollinator is an animal that moves pollen from the male anther of a flower to the female stigma of a flower. ... A pollinator is different from a pollenizer, a plant that is a source of pollen for the pollination process

Philosophies – the study of the fundamental nature of knowledge, reality, and existence, especially when considered as an academic discipline

Poaching – Poaching has been defined as the illegal hunting or capturing of wild animals, usually associated with land use rights

Pejorative – expressing contempt or disapproval

Prophesies – say that (a specified thing) will happen in the future

Paradigm - a typical example or pattern of something

R

Realm – a kingdom

Repository - a place where or receptacle in which things are or may be stored

Rodent – a gnawing mammal of an order that includes rats, mice, squirrels, hamsters, porcupines, and their relatives, distinguished by strong constantly growing incisors and no canine teeth. They constitute the largest order of mammals

Rustling – make a soft, muffled crackling sound like that caused by the movement of dry leaves or paper.

Revelation – a surprising and previously unknown fact that has been disclosed to others

Repudiate - refuse to accept

Reservoir - a large natural or artificial lake used as a source of water supply

Reincarnation – the rebirth of a soul in another body

Rafting – the sport or pastime of travelling down a river on a raft

Recreation – activity done for enjoyment when one is not working

Radiation - the emission of energy as electromagnetic waves or as moving subatomic particles, especially high-energy particles which cause ionization

S

Superstition - excessively credulous belief in and reverence for the supernatural

Sacrifice - an act of slaughtering an animal or person or surrendering a possession as an offering to a deity

Sacrament – (in the Christian Church) a religious ceremony or ritual regarded as imparting divine grace, such as baptism, the Eucharist and (in the Roman Catholic and many Orthodox Churches) penance and the anointing of the sick

Scriptures - the sacred writings of Christianity contained in the Bible

Scintillating – sparkling or shining brightly

Scoping - an investigation of wide scope

Symposium – a conference or meeting to discuss a particular subject

Sovereign - a supreme ruler, especially a monarch

Sociologists – an expert in or student of the development, structure, and functioning of human society

Solidarity – unity or agreement of feeling or action, especially among individuals with a common interest; mutual support within a group

Sustenance - food and drink regarded as a source of strength; nourishment

T

Twinkling – shining with a gleam that changes from bright to faint

Torment – severe physical or mental suffering

Trafficking – deal or trade in something illegal

Treatise – a written work dealing formally and systematically with a subject

Trample – tread on and crush

Toxins - a poison of plant or animal origin, especially one produced by or derived from microorganisms and acting as an antigen in the body

U

Upanishads - each of a series of Hindu sacred treatises written in Sanskrit c. 800–200 BC, expounding the Vedas in predominantly mystical and monistic terms

Unanimously – without opposition; with the agreement of all people involved

Unequivocally – in a way that leaves no doubt

V

Veneration - great respect; reverence

LIST OF FIGURES

PREFACE

This book is about two main pillars of human existence. One is 'Religion' and the other one is the 'Environment'. Religion is the most debatable topic in the world and a sensitive topic that must be used with care. The environment is known to all but the essence of it is not really understood by many. This book deals with these two important topics and it also analyzes the way in which both are connected.

The book is designed in a way that it enhances the spirituality and divinity among the readers and to look at religion is an absolutely beautiful way. The initial Chapter focuses only on religion and many unknowing facts about various religions are also explained. There is more information shared about the different religions in the world, the principles, and teachings of all those religions.

This book will take the readers through the fundamentals of religion and explain how religion helps in transforming an individual into a better human being. The book will also give readers the answer to the origin of religion and the history behind religion. It will explain how religion helped in the development of men and society.

The next aspect discussed in the book is about the environment. There is a separate Chapter dedicated to 'Nature' and this chapter emphasizes that nature is the biggest of all religions. This Chapter gives the readers valuable inputs that they can follow in their life. The importance of the environment and various other things that the environment provides us are also talked about.

The beautiful connection between the two life-changing forces i.e., religion and environment can be understood by reading this book. How religion teaches us to be environmentally friendly and how the environment is connected with our human body is also explained in detail. The book will make the readers understand how well religion and environment complement each other. The book also iterates that religion is pure and it should be taken in that sense. Superstitious beliefs, unwanted sacrifices, and myths in the name of religion must be avoided.

Everything in the world has its positives and negatives or we can rather say that everything in this world can be used in an efficient way or inefficient way.

So, we will also discuss how religion and environment are used inefficiently by humans, what are its consequences and how to solve the problems. Environment awareness should be more important that religious awareness for people of today. Hence a separate Chapter is allocated to analyze the different environmental issues that our planet faces and what we, as residents of this planet can do to solve them.

This book is created with the intention that both religion and environment are equally important for the well-being of individuals as well as for the development of society. Hope the readers enjoy the book and use the knowledge gained in a constructive way.

What is Religion?

1

"Religion is nothing else but love to God and Man."

William Penn

CONTENTS

'Religion' is a very complicated term and it must be used with great care. The word religion is in existence from quite a long time. There are many people who blindly do anything and everything for religion and many who do not believe in this concept also. Many people fear to talk about this subject because it involves the supreme power 'God' and the beliefs based on God. There are many religions in this world and each religion has its own followers. This Chapter serves as an introduction and will explicitly define what religion is and why religion exists. We will also learn about different religions existing in this world and some important facts about each religion.

1.1 DEFINITION AND UNDERSTANDING OF RELIGION

1.1.1 Definition

'Religion' is something that all human beings speak about either positively or negatively. There are some sections of people who even go to the extent of sacrificing their lives for the sake of religion and there are some group of people who says that there is no religion. Although the difference of opinion persists, religion is widespread across the globe in different forms. It is very difficult to define the term 'Religion' because it involves people's faith which is very sensitive. Many scholars and legends have tried to define religion. While few succeeded to some extent, most of them have failed in their attempt. Even the definitions framed were thought to be based on some specific region or specific people. William James, one of the highly respected surveyors of religion is supportive to the fact that "the man who knows religion most completely troubles himself least about a definition". Still, defining religion will help a layman to understand the term in a better way.

According to dictionary, "A religion is a particular system of belief in a god or gods and the activities that are connected with this system."

The word 'Religion' is said to have its origin from the Latin word 'Rel (I) igio' which has two root words. The first root word meant belief in and practice of 'signs of Divine Communication'. The second root word refers to the activities that link human beings with supernatural powers. So, belief and practices based on the beliefs are the main characteristics of all religions.

Figure 1.1: Definition of Religion.

Source: https://www.youtube.com/watch?v=wIzdlrZI2Mg

Whenever someone tries to define religion, it invokes controversy because they try to define God. In today's technological world, even the scientists agree that there is certainly something more powerful which is above human's intellect, still, there are many unanswered questions in this scientific world.

Many great legends and scholars have tried to define religion from their experience and understanding. Let us see few among them:-

- In Pfleiderer's words, religion is "that reference men's life to a word governing power which seeks to grow into a living union with it."
- According to Arnold W. Green, religion is "a system of beliefs and symbolic practices and objects, governed by faith rather than by knowledge, which relates man to an unseen supernatural realm beyond the known and beyond the controllable."
- As put forth by Gillin, "The social field of religion may be regarded as including those emotionalized beliefs prevalent in a social group concurring the supernatural plus crest and behavior, material objects and symbols associated with such beliefs."
- MacIver and Page defined religion as "Religion, as we understand the term, implies a relationship not merely between man and man but also between man and some higher power."
- James G. Frazer considered religion, "to be a belief in Powers superior to man which are believed to direct and control the course of nature and of human life".
- Sumner and Keller were of the opinion that, "Religion in history, from the earliest to very recent days, has not been a matter of morality at all but of rites, rituals, observance and ceremony".
- According to Christopher Dauson, "Whenever and wherever man has a sense of dependence on external powers which are conceived as mysterious and higher than man's own, there is religion, and the feelings of awe and self-abasement with which man is filled in the presence of such powers is essentially a religious emotion, the root of worship and prayer."

Figure 1.2: Definition of Religions by Legends and Scholars.

There are numerous definitions that explain religion. As we know, religion by itself has different forms and beliefs and so the way it is explained is also infinite in number. Each pick any one definition based on their belief and what they practice. It is through religion that humans preserve the values of life. Religion cannot always be defined based on any supernatural power or God because there are also religions that do not worship any God like Buddhism.

Religion is a universal thing and it exists for a very long time. We can even say that the base of culture is religion. Religion is a belief which people have on a supreme power or being and they have certain activities and behaviors based on their belief which have transformed into culture. Even the festivals that people celebrate are based on their religions. There are many epics and stories behind every ritual and every celebration. It is the faith upon a particular religion that binds humans together. All members of a particular religion share a common belief and share common practices.

Figure 1.3: Different Religious Practices.

Source: http://www.yourarticlelibrary.com/religion/essay-on-religion-meaning-nature-role-and-other-details-5931-words/8586

Most of the religions are attached to a specific God or an idol and these religions have their own identities and religious practices. But people today are changing the way they look at religion. For example, there are some Christians who say that Christianity is not a religion but a relationship with Jesus; similarly, some Muslims say that Islam is not a religion and is a way of life. There is a growing trend now that many are terming their belief to be a 'way of life' and not religion.

People use religion as a gateway to seek God. Great saints, priests and monks acquire scholarly knowledge and learn the essence of life because they practice what they preach. Religion has been in existence during the past, it exists today and it will exist even in the future. There will be no society without religion. People hold to their religions during their joyful moments as well as during a crisis. Hinduism, Christianity, Islam, Judaism, Buddhism, Confucianism, Taoism, and Shinto, etc., are some of the greatest religions in this world.

1.1.2 Nature

There are many religions practiced by people all over the world. When we look upon religions as a whole, we can very well see that every religious people have some common characteristics and their attitude towards life will mostly be the same. Every religion is unique in its belief systems, method of worship, rituals and rites, sacrament, interdicts, moral prescription and every religion own their own sacred books also.

Figure 1.4: Religion.

Source: http://www.englishspectrum.com/d-6-visa-religious-works/religious-words-25-1024x683/

Radin states that every religion consists of two parts – Physiological attachment and Psychological attachment. The Physiological attachment means the religious practices done physically such as praying by closing the eyes, touching the feet, kneeling down and such. The second one is the attachment one has towards their religious beliefs and traditions.

If all the ancient and great religions of the world were analyzed, we can certainly find some basic elements that every religion has in it. Let's look into detail what these elements are:

The Existence of a Supreme Power

Figure 1.5: Picture Representing an Unknown Supreme Power.

Source: https://www.express.co.uk/news/science/756870/proof-of-god-kurt-godel

If a person follows a certain religion, it means that he believes that there is some power above us which has created us and is shaping us. They believe that it is that supernatural force which has created the Mother Nature and that every movement of even the tiniest atom in this universe is designed by this force. They literally believe that although Thomas Alva Edison invented light, the knowledge to him was given by this supreme force. Some people call this force as God and some others refer to it as some external supernatural force in the universe.

Rituals

There is no religion without rituals. Rituals are certain kind of procedures followed in a religion which is unique to that religion. These rituals will be followed hereditary by the members of a particular religion; these rituals will have strong meaning and if we analyze very deeply to find its root, it would have been preached by some great sages, rishis, priests based on the Upanishads, holy scriptures, etc.,

Figure 1.6: A Ritual Performed by Buddhists.

Source:https://commons.wikimedia.org/wiki/File:Yee_Peng_Sansai_Thailand,_Buddhist_culture_religion_rites_rituals_sights.jpg

Rituals are the traditional part of the religion. It is the tradition followed in many religions to celebrate a particular day or a particular month for different reasons. Even celebrations at home like the naming ceremony of a child, birthday celebration according to date or birth star, marriage, anniversaries and even mourning is practiced differently in different religions.

These rituals exist from ancient times and there are nces pieces of evidence which prove this fact. Scientists have discovered that Neanderthal people started burying their dead some 120,000 years ago. Early *Homo sapiens* put tools and other things into graves with the bodies, as if they could use them in the afterlife. The artworks and many small objects found in the graves w was were dated 40,000 years ago was believed to be placed in the graves for religious reasons.

It is these rituals that keep the religion active and attract more followers. These rituals are not a one-time affair; they are practiced repeatedly as part of daily life. Few examples of rituals are singing, dancing, reciting prayers, hymns, crawling, fasting, feasting, weeping, etc., A person who follows a particular religion is expected to follow the rituals that are linked to that religion.

Respecting Holy Things

In every religion, there are certain things which are considered to be holy or sacred. And every follower of that religion respects those holy things. For example, the water of River Ganges is considered holy by Hindus and they believe that getting a dip in the Ganges or even sprinkling some water from river Ganges would wash away their sins. Hindus even worship cow and cow is considered very auspicious and a good omen. In Hinduism, some animals represent the different incarnation of Gods and so these animals are also given the same respect as that of the supernatural force called 'God'.

Figure 1.7: Hindu Devotees Taking Dip in the Holy Ganges River.

Source: http://www.bluebird-electric.net/oceanography/Rivers/The_River_Ganges.htm

Sinful Activities

Every religion has defined certain activities as sinful activities and it is believed that following these activities will cause harm to the person and their family. People belonging to every religion strongly believe that any sinful activities caused by humans will displease God and they may have to face the wrath of the supreme force. There are also some beliefs in a few religions that if a person commits sin then he may have to take birth again and will face the consequences of the bad karma.

Figure 1.8: Seven Deadly Sins as Per Christianity.

Source: https://www.pinterest.com/pin/258253359860518571/

Path to Salvation

Every religion preaches its own method of salvation. The term 'Salvation' denotes the ways by which a person rectifies his guilt, pulls him out of sinful activities and reaches the feet of God. In Hindu religion, salvation is referred to as 'moksha' and according to Hinduism, the purpose of life is to attain moksha which means removing all the sinful acts and bad karmas in this birth and reaching God after death. As per Buddhism, salvation means being absorbed in the Godhead and attaining Nirvana. Christians believe that their Lord Jesus Christ himself has given his life for man's sins.

Figure 1.9: Attaining Nirvana or Enlightenment, the Path of Salvation in Buddhism.

Source: https://successguruakmishra.wordpress.com/2015/10/26/concept-of-salvation-in-different-religions/

Holy Book

Every religion has its own holy book and different religions call it by different names. Usually, the teachings of a religion are written in the holy book. It may be in a story form or as commandments. The form of writing may be different but the very essence of every holy book is to help people understand the meaning of life and the basic principles that must be followed to live a happy and contented life.

In short, every religion has a set of beliefs and some elements that they follow to get nearer to the supreme force or the divine force. The ultimate goal of every religion is to help mankind and to lead them in the right path. No human being in this world is perfect. Men believe that following the principles of their religion will please God and that will help them to get a place in heaven (considered as the best salvation).

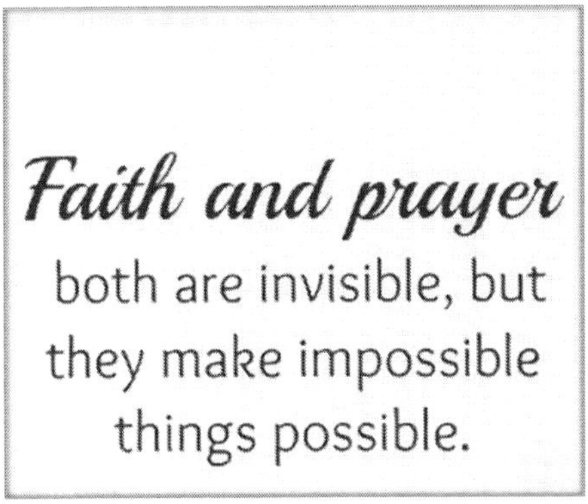

Figure 1.11: Why Does One Pray?

Source: https://me.me/i/faith-adl-prayer-both-are-invisible-but-they-make-impossible-16660574

In religions where no God is worshipped, the followers rely on their guru's teachings to find the answer to their problem. For example, in Buddhism, there is no God worship instead they rely totally on Buddha's teachings. There are many ways in which religion help mankind. Without religion, humans will find it very difficult to live in this vast universe.

It is the religion that teaches the most important morals of life.

"Morals" are the way by which one human behaves with other humans.

The existence of the mighty force and how the mighty force wants humans to live and behave forms the basic essence of religion and this is also the main reason why we need religion.

Every religion has certain rules as to how humans must behave with one another and with other living creatures. 'Tao' in China means following a "path" of goodness, truth and duty and adapting to this is very important in some religions. In Judaism, the moral taught to people is "love your neighbor as yourself". Christianity urges people to think of every single person as their "neighbor" and treat them with love. Although most of the religions teach people to behave kindly with others, there are some religions where people believe that they can please God by killing or sacrificing another person.

Religion teaches every aspect of life. For example, in Hinduism, great scholars admit that for every problem man faces in life, if he needs a solution, it can be found in the ancient Vedic scriptures. Even before people developed reasoning skills, religion has been the chief supporter of the spiritual and moral values of life. Religion plays an important role in economics and politics also. The contribution made by religion to the society is vast and it is because of these contributions that religion is much needed.

Let us discuss in detail some of the functions attributedreligion to the society:

1.2.2 Contributions of Religion

Religion Develops Helping Nature.

Every religion motivates its followers to help the poor and needy. Most of the religions in his world states that by helping the needy one can please the supreme power called God and attain his blessings.

Figure 1.12: All Major Religions Teach Charity.

Source: https://lifehacker.com/seven-important-lessons-from-world-religions-everyone-s-1613615832

Tr followers of a religion follow whatever is taught to them as religious principles and this helps develop philanthropic nature in people. Even today we can see free food being provided to people at temples, churches and other religious institutions. Thus, religion does not stop with bringing good out of people but also using to improve the society as a whole. When people start sharing, the thought of discrimination diminishes. The religious leaders and priests were the people who started providing medicine for the sick. This is from where science developed. They inculcated the habit of charity among people and soon kind-hearted rich people started to follow their path. This resulted in the opening of many charitable institutions like hospitals, health centers, rest houses, temples, etc.,

Religion Unifies People in a Unique Way

Religion is certainly the unifying force that binds everyone together. Every religion has com values that is are meant for all and followers of each religion unanimously follow these. Common faith, values, and norms etc. are significant in unifying people. Rituals are common in every religion and people individually or collectively practice these rituals. When people collectively do things together, their relationship improves and they start to love being together.

Figure 1.13: Devotees Performing a Ritual in Group.

Source: http://creative.sulekha.com/deepavali-greetings_620391_blog

Religion improves integration within the group members. It regulates the conduct of individuals by enforcing moral principles on them and by prescribing powerful sanctions against them for violation.

Religion Makes Survival Possible

Historians say that religion has played a crucial role in the development of the early civilizations such as Indian civilization, Chinese, Egyptian Mesopotamian, Cretan, Middle American and Andean. When everything was new and survival seemed impossible, it was the religion and religious believes thae people the much much-needed courage and knowledge to sustain.

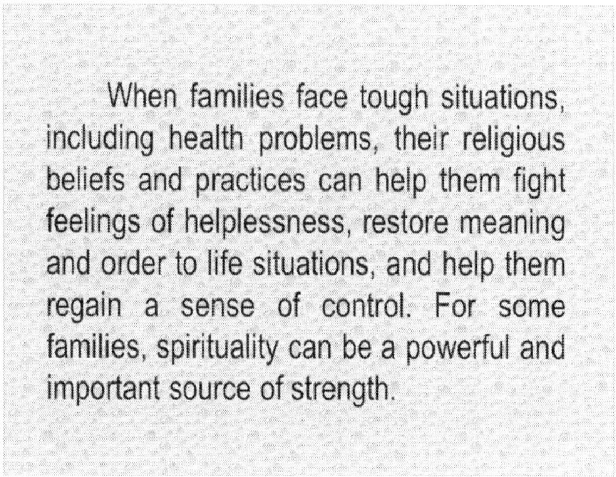

Figure 1.14: Power of Spirituality.

Source: https://slideplayer.comslide/8933495/

RReligion provides an explanation to handle tough situations in life in unfavorable situations. Religion gives a sense of security which cannot be provided by anything else in this world. This sense of security of the individual has greagnificance in the society. Since religion helps man to forget the suffering, disappointments, and sorrows in this life, social dissatisfaction and social unrest become less frequent and the social system continues functioning.

Religion Develops Relationships

It is due to the existence of religions that families flourish. Religion shapes up people and their relationship especially marriage and family. There are many rituals performed in every religion that brings the entire family together and make them understand the value of togetherness.

Figure 1.15: Image of a Family Praying Together.

Source: https://askgramps.org/why-should-one-kneel-while-family-prayer-is-being-said/

Humans started giving importance to themselves only because of religion. They feel proud of themselves because they are the most wonderful creation of God. They feel triumphant since the purpose of their creation is to know about God and to unite with him. Religion is the primary reason for the development of civilization.

Religion Teaches Value

Religion teaches the important values of life. It is the values that control every action of a human being. Values of life are based on religious believes. Religion is the foundation upon which these values rest. Obeying and respecting the parents, always speaking the truth, being sincere and faithful in a relationship, honesty, being humble and modest, etc., are some of the good values preached by religion which are important both to individuals and to the society as a whole. Possession of good values is the pmary requirement of a society. Every religion preaches love, sacrifice, and non-violence. All religions ask men to get rid of the desire to live a happy life.

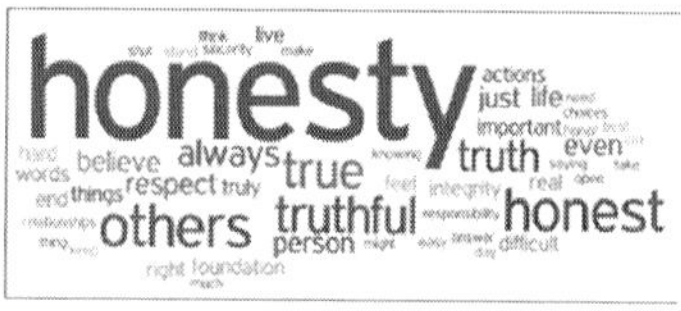

Figure 1.16: Values Taught by Most Religions.

Source: http://www.st-marys-purton.wilts.sch.uk/our-curriculum/worship-themes

Religion Controls the Society

Figure 1.17: Religion as a Way of Social Control.

Source: https://www.slideshare.net/FaimaMoHammad/social-control-and-agencies-of-social-control

The social behaviors practiced by every individual in a society is controlled by religions. Certain social behaviors are considered nsive not only in the society but they are also considered as a sin against God. People fear the supernatural force and this feara reason for adapting adopting the right social behavior. Religion not only defines moral expectations for members of the religious group but usually enforces them. It supports certain types of social conduct by placing the powerful sanctions of the supernatural behind them. Any violation of the acceptable norm is punishable not only by God but by society. Hinduism gives sanction to the caste system which regulates social relations of various classes in India.

Religion is an Energy Booster

In this world, human beings are subjected to many sufferings and disappointments. We often hear news about people committing suicide due to depression. There are many times where even the last ray of hope vanishes; sometimes what we plan collapses right in front of our eyes. We feel like we are lost and do not know what to do next. In such situations, every human being needs some sort of consolation and someone to rely on. It is here where religion plays an important role. The faith and belief that ns have on religion has have enormous power to pull them out of grief or any miserable situations.

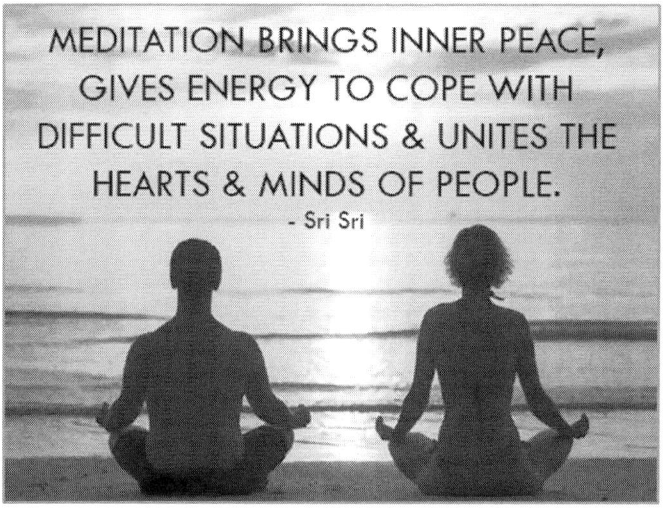

Figure 1.18: A Religious Practice That Gives Immense Energy Physically & Mentally

Source: http://goholistic.com.au/?p=7073

Religion says that even the sufferings are given to humans to make them stronor future tasks that is are bestowed upon them. It is the faith that humans have on hat gives them the much much-needed interest to face life again and makes all the pain bearable. In this way, religion helps man to bear his frustrations and encourages him to accept his lot on earth.

Religion has not only proven to discipline mankind but it has also exerted a great influence over economic and political life. Even today religion is calld upon to support rulers, and other legal procedures.

1.3 RELIGIONS OF THE WORLD

Although there is a general belief that God is one and only the forms are different, people worship their own form of God under various religious names. Since the concept of religion came into existence a long time ago, as the years passed by numerous other religions popped up. According to Adherents, an independent, non-religiously affiliated organization that monitors the number and size of the world's religions, there are approximately 4300 religions in the world. Every religion has different beliefs and princ each religion has their its own way of worship and rules; the form of rituals also differs from religion to religion.

Figure 1.19: Religions of the World.

Source: https://justsimplyinlove.wordpress.com/2012/07/20/fasting-in-different-religions/religions/

There are also religions that share common belief and practices. Followers of religion can be called as 'believers', or 'the faithful'. Few people follow more than one religion at a time. Hinduism, Christianity, Islam, Buddhism, Taoism, Sikhism, Judaism and Jainism are some of the largest religions of the worlch of these religions have has followers across the world.

Let us learn about some of the oldest and most widespread religions in the world:

1.3.1 Hinduism

Hinduism is a very ancient religion and it is the third largest religion after

Christianity and Islam with more than 1 billion followers worldwide. Hinduism has its base in India and is practiced by most of the Indians. The followers of Hinduism are called Hindus. Almost 80% of the people living in India are Hindus. It is said that the word Hindu is derived from the Sanskrit word 'Sindhu' which meant 'river' and in this case denotes "of the Indus Valley" or "Indian".

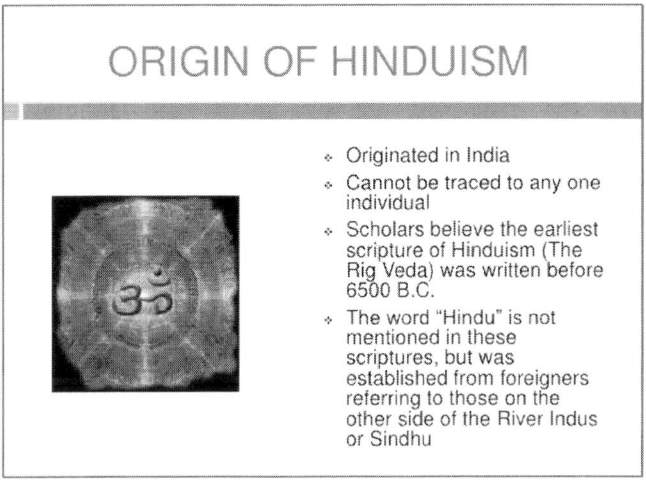

Figure 1.20: Origin of Hinduism.

Source: https://www.slideshare.net/jbstubb77/ap-human-geography-2011-hinduism

There is no historical evidence to show the founder of Hinduism or to state its date of origin. But the old Hindu scriptures are dated to be of 1500 – 1200 BCE which makes this religion to be one of the oldest surviving religion in the world. Although Hinduism may look complex from outside, the beauty and the divinity of this religion can be known only if one understands the true beliefs and principles of Hinduism. No religion can explain the beauty of life and reason for our sufferings as candidly as Hinduism. It is not all; no religion other than Hinduism can help you to get detached from the materialistic life and still remain happy.

Hinduism is not a homogeneous religion. Hindus worship multiple Gods. Most of the Gods worshipped by Hindus represent different forms of nature. This is one of the significances of this religion.

Figure 1.21: Meenakshi Amman Temple, A Place of Worship of Hindus.

Source: https://www.nationalgeographic.com/travel/destinations/asia/india/meenakshi-amman-hindu-temple/

Hinduism is always connected with nature; it gives importance to nature and it values nature. Almost all the rituals practiced in Hinduism will indulge any of the elements of nature for sure. Basically, Hinduism proves to the world that mother nature is the reason for our survival and we have to live along with nature without disturbing it.

Figure 1.22: Different Forms of Gods Worshipped by Hindus.

Source: https://twitter.com/hinduismfromw

Hinduism started to grow more philosophically with the spread of Upanishads and Vedas around 700 BCE. The Vedas and Upanishads are the sacred texts where one can know about God and what God teaches us and these are treated with high respect. Hindus worship three main Gods – the Brahma (the Creator), Vishnu (the Protector) and Shiva (the destroyer of evils). As Brahma is common to all, some Hindus are strong believers of Vishnu and some are devotees of Shiva. There also many other Gods and each are is connected to others in different ways.

Figure 1.23: Image of Hindu Gods Brahma, Vishnu and Shiva.

Source: https://www.templepurohit.com/origin-lord-shiva-lord-vishnu/

The true followers of Hinduism are very spiritual. Worshipping God and performing their family duties tops the priority lists of every Hindu. They fear God and they believe that their current sufferings are due to their deeds in their previous birth and what they do now will determine their future in this birth as well as in the next birth. Hindus are strong believers of Karma and Moksha and the concept of Karma has come from Hinduism. What Karma means is that a person faces Pain, disease, poverty etc., because of their own evil actions, usually from a previous lifetime. The goal of every Hindu is to release themselves from their Karma and attain moksha in this birth itself which means if one does not do any sin in his lifetime and be good to all, he will not have any re-birth and can reach God.

Hinduism has so many classifications and even the followers are categorized under various sub-divisions based on the nature of their work. Given below are some of the main divisions:

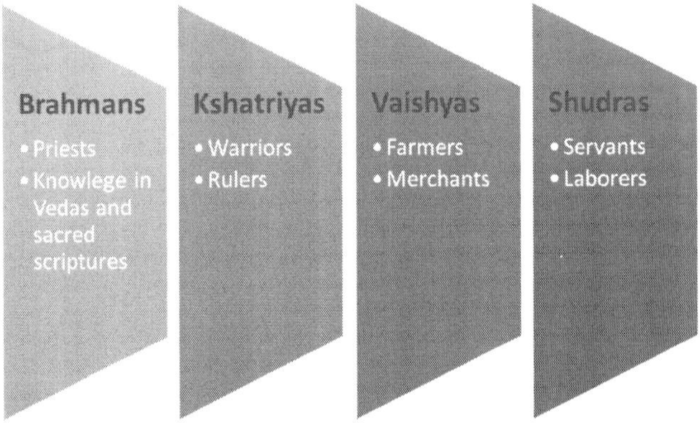

Figure 1.24: People Classified Under HinduisBrahmans are considered to be the high high-class people and they are highly respected. Shudras on the other side are considered to be the low class. These are basically the divisions created by man based on the qualification and job nature of people; this reference to classification has been mentioned in the great epics like Ramayana and Mahabharata.

Even though there exist such classifications, Hinduism gives importance to Dharma which means Justice. Hinduism is based on the ideas of Dharma. Dharma denotes the universal rules that could help people become happy and avoid suffering. It also means the discharge of duties.

The belief sys and the morals followed by Hinduism is are so exceptional that few other religions started sprouting out from Hinduism around 500 BCE namely Buddhism and Jainism. Hinduism began to gain popularity in the West during the 20th Century. What has attracted westerners the most is the belief system, the culture and the tradition of the Hindu religion. They understand that Hinduism does not only provide spiritual well-being but it also provides mental well-being and good health by way of meditation and yoga.

Figure 1.25: Westerners Following Hindu Customs.

Source: https://postcard.news/westerners-getting-attracted-hinduism/

Figure 1.26: Yoga, a Hindu Spiritual Practice.

Source: http://vishalmaliwal.com/blog/yoga-yoga-information-yoga-techniques-yogas-types/

There are various religious movements like Hare Krishna and New Age, and even more so through the incorporation of Indian beliefs and practices (such as the chakra system and yoga). To say honestly, Yoga is one of the best things that the world has derived from Hinduism and no one can deny it. The culture followed by Hindus is very popular in Europe and America. The holy text of Hindus is the Bhagavad Gita and there are also other sacred books such as the Vedas and the Upanishads that can provide an answer to any of our life problems.

Figure 1.27: The Bhagavad Gita.

Source: http://krsnastore.com/srila-prabhupada-books/3-pocket-edition-bhagavad-gita-as-it-is-original-edition-kbi-9781602930124.html

1.3.2 Buddhism

The religion Buddhism is well-known for the discipline that it teaches its followers, the moral values and the directives that a person may want to live by. Buddhism dominates the eastern part of the world and it has spread tremendously. There are over 360 million followers of Buddhism worldwide and over a million American Buddhists today. The followers of Buddhism are called Buddhists and they do not worship any God and they do not believe in any God. The sign denoting Buddhism which is a universal sign of nill, a circle with a slanted line through it is to illustrate that Buddhists do not believe in any God.

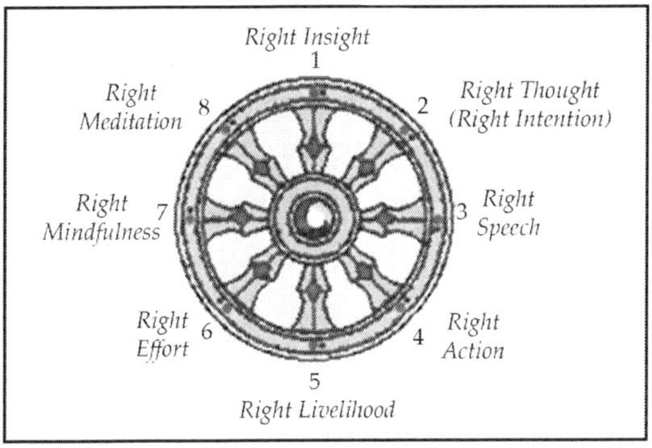

Figure 1.28: Sign of Buddhism and What It Denotes.

Source: https://rst61-2spring2015groupl.weebly.com/beliefs-and-practices4.html

Often it is misunderstood by the outside world that the Buddhists worship Gautama Buddha who is the founder of Buddhism. But the truth is that Gautama Buddha never claimed him to be a God or a person with divine powers and nor the followers worship him as God. The Buddhists try to walk in the footsteps of Gautama Buddha by following his teachings and attain spiritual enlightenment similar to him.

Figure 1.29: Siddharta Gautama Attained Enlightenment Under Bodhi Tree.

Source: https://www.buddhist-art.net/about-buddha/

Siddharta Gautama was a Nepali Prince who lived around 500 BCE and he led a luxurious life. Unable to bear the sufferings of other people outside his palace walls, he abandoned his luxurious life and went in seek of answers for this discrimination and sufferings of life and how to relieve oneself from this suffering. Finally, one day, he got enlightenment and found the answers to all his questions and he became the Enlightened One – the Buddha. He spent the remaining 45 years of his life teaching the dharma -the path to liberation from suffering and establishing the sangha which is a community of monks.

The teachings of Buddha are the sole principles of Buddhism. Buddhists follow a list of religious principles and adhere to personal restraint, fasting and very dedicated meditation. Similar to Hinduism, the ending of re-births is the ultimate goal of every Buddhist because they too believe that rebirth means suffering. According to Buddhism, it is the desire, aversion, and delusion that urges man to commit sin and leads to re-birth. Purification of heart and leaving off all the worldly desires is the main goal of Buddhism. This state is described in Buddhism as "Nirvana" – meaning blowing out of the flame of desire and they achieve it through sheer meditation. The main teaching of Buddhism is that nothing is permanent in this world.

Figure 1.30: Meaning of Nirvana in Buddhism.

Source: https://spiritualray.com/what-does-nirvana-mean-in-buddhism

Meditation in Buddhism has more importance than prayer and focusing on God. Buddhism stresses more on self-discipline. There are different beliefs in Buddhism; some follow rituals and some favor pure meditation. Whatever be the mode, they all have one thing in common which is the respect they have on the teachings of Buddha.

Figure 1.31: Types of Buddhism.

Theravada Buddhism and Mahayana Buddhism are some forms of Buddhism. Theravada Buddhism which is atheistic and philosophical in nature is prominently found in South-east Asia; this form focuses on the monastic life and meditation as a means to liberation. Quite alternatively, Mahayana Buddhism practices rituals or meditation or the combination of both. This form is practiced in China and Japan and it incorporates several deities, celestial beings, and other traditional religious elements. Zen, Nichiren, Tendai, and Pure Land are some of the major forms of Mahayana Buddhism.

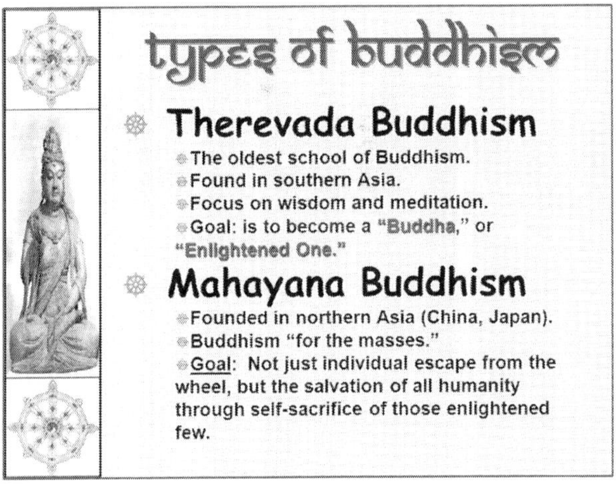

Figure 1.32: Important Facts About Therevada and Mahayana Buddhism.

Source: https://slideplayer.com/slide/4264612/

Buddhism is popular among westerners as a religion with importance to meditation. The important element in Buddhism is meditatioare two main types of Bditation – Vipissana Vipassana and Samatha. Vipissana Vipassana is about understanding the reasons for suffering in the world and how to end it and Samatha meditation is about tranquility and love.

The Tripitaka is the holy book of the Buddhists. The book is also called the Pali Canon, named after the language in which it was first written. The book contains the oral teachings of Buddha.

Figure 1.33: Tripitaka, the Sacred Book of Buddhism.

Source: https://www.amazon.in/Tripitaka-Buddhism-Anil-Kumar-Singh/dp/8188540684

1.3.3 Christianity

Christianity is one of the largest religions in the world with more than 2 million adherents. The followers of Christianity are called Christians and they worship Jesus Christ as their Lord. Unlike Buddha, Jesus Christ was

According to Christians, all human beings have committed sin and the punishment for sin is death. Jesus, out of his love for humans came in human form, suffered all the pain and died willingly on the cross repaying for all the sin committed by humanity. Jesus who died on the cross rose again after three days as he promised to prove his divinity.

This belief of Christianity proves that God immensely forgives humans for their sins. That is why he has sent his Son Jesus to the earth to pay for all our sins. The faith that Christians have on Jesus is unshakable because it was Jesus who said that he would be crucified on a cross and that after three days of his crucifixion, he would come back to life and the same happened. On the third day, Jesus' tomb was found empty and many people testified that they saw him alive again.

Christians believe that Jesus Christ is the ultimate truth and they follow him and his teachings. 'Holy Bible' is the sacred book of Christians and each and every follower of Jesus believes that the Bible has the message from God himself to mankind.

Figure 1.37: The Holy Bible, Sacred Book of Christians.

Source: https://www.shutterstock.com/video/clip-943537-holy-bible-glowing-rays-background

The Bible is looked as a treasure and it contains the historical record of Jesus' life and his miracles. By reading a Bible, one can know more about Jesus, his personality, his love for humans and how can one relate to God as a friend. Jesus is seen as a loving God who is always merciful to all living beings.

1.3.4 Islam

Islam is the second largest religion next to Christianity with more than 1 million followers across the world. Islam is largely widespread in the Arabian countries. There is a general assumption all believers of Islam are Arabs but the real fact is that more more numbers of Islamic people live outside the Arabic nations of the Middle East. The followers of Islam are called Muslims. Muslims believe that there is only one God named 'Allah' who is the supreme commander of mankind. According to Islam, Allah is the creator of the universe and whatever happens is as per his will. A devotee's relationship with Allah is seen as a relationship of a slave to his master.

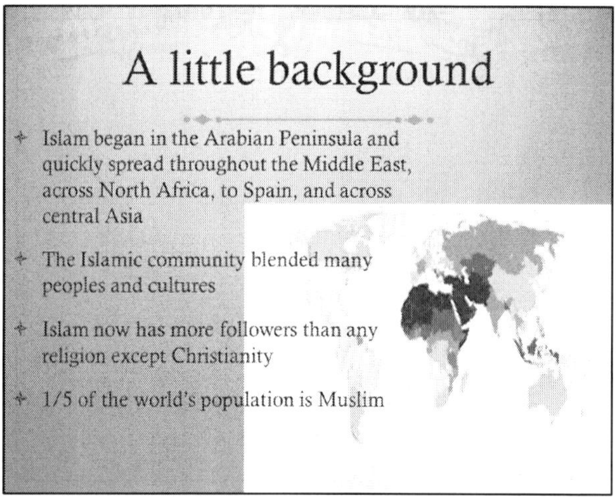

Figure 1.38: Background Information About Islam.

Source: https://slideplayer.com/slide/5768299/

Similar to all other religions, even Muslims believe that a person is rewarded or punished after death based on the good and bad deeds that he performs during his lifetime. Doing good deeds ensures one's entry into Paradise after death and giving up one's life for Allah is considered as a sure way to enter Paradise. The Quran is the holy book of Islam. The Quran reveals the nature of God to the Muslims. The Quran is regarded as a guide to living life in harmony with the Creator, Allah

Figure 1.39: The Quaran, Holy book of Islam.

Source: https://www.reference.com/world-view/holy-book-islam-called-e8e735e28094ce3e

Muslims also honor Muhammad who is the last prophet. Prophets are people who saw things that a normal human being could not see and told people what they should do to enjoy life after death and save themselves from troubles after death.

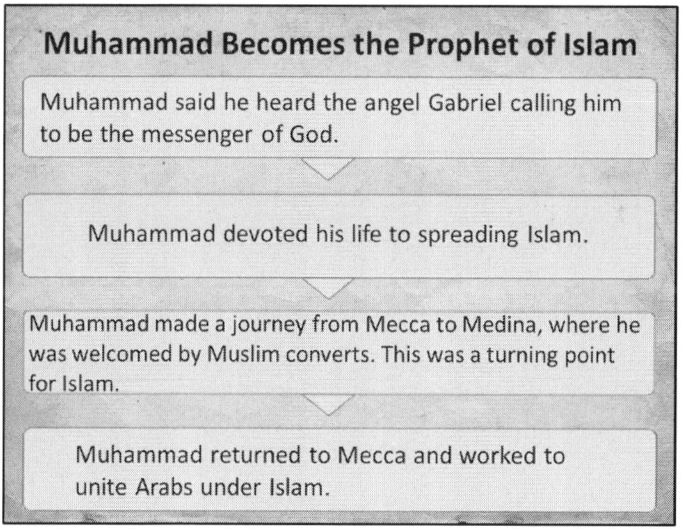

Figure 1.40: Muhammad, the Prophet of Islam.

Source: https://slideplayer.com/e/8879233/

A person following the Islamic religion must lead a verymuch disciplined life. Each and every Muslim, no matter where in the corner of the world they are, they must follow the below mentioned five religious duties:

- Repeat a creed about Allah and Muhammad
- Must recite certain prayers in Arabic five times a day without fail. The Quran is the holy scripture of Muslims.
- Help the needy
- Every year, Muslims must fast from food, drink, sex and smoking from sunrise to sunset for one month
- Must take a pilgrimage to Mecca to worship the shrine at least once in their lifetime

Figure 1.41: Five Pillars of Islam.

According to the Islamic faith, a true Muslim is one who performs all of these duties and they believe that if one performs all these duties, they can enter the heaven and people who fail to do this will be punished in hell. Muslim countries are known for their rigidness. Muslims take their religion and principles of their religion very seriously. They never compromise on their religion and its beliefs. If anyone questions the Quran in Islamic countries, they are liable to be killed. Muslim clerics can still tolerate questioning of Allah, but if their Prophet is questioned, they start proclaiming fatwa of death for the questioner.

There are also many other religions in the world such as Confucianism/Taoism, Shintoism, Sikhism, Judaism, Zoroastrianism, Bahai etc., The below table gives a rough figure on the number of followers of these religions:

BELIEFS	NUMBER OF FOLLOWERS
Confucianism/Taoism	540 million
Shintoism	110 million
Sikhism	19 million
Judaism	14 million
Zoroastrianism	150,000

Figure 1.42: List of Other Religions.

Each of the above religions has their own set of beliefs, modes of worship and other spiritual practices. Each has further sub-sections. For example, in Sikhism, we have Namdharis, Nirankaris, Udasis etc., and in Jainism, we have Shwetambar, Digambars and various sub-sects. People in different corners of the world might have different faiths but all these religions exist

because most humans believe that there is a unique power that has created us and guiding us in all walks of life.

1.3.5 Commonality Among Religions

We learned about some of the major religions in the above section. Although each religion is given various names and has its own principles and believes, there are many things that are common among most of the religions. Diversified religions have some common agenda.

Let us look into some of the common practices and beliefs followed by people under various names:

- Every religion believes in an unknown power i.e., God and each religion call him by giving different names. Most religions also believe in the existence of heaven and hell. Good doers are sent to heaven where God resides and evil doers are sent to hell for punishment.
- Most religions believe in the transformation of souls which means when a person dies, it the body that dies and not the soul. The soul still remains and takes re-birth or reaches the almighty based on the deeds that the human has done in his present life.
- Theory of Karma is widely accepted by almost all the religions but with different name tags and slight variations. On a general note, the Karma theory meant our present situations are supposed to be the result of our past deeds and our future situations would be the result of our present deeds. So, almost all religions insist on performing good things as it is the key to a happy life.
- All the religions in the world preacs good qualities like charity, forgiveness, patience, purity of heart, self-control, being true, humanity, kindness, and free from anger. More importantly, all religions insist on non-violence. To achieve most of these qualities, meditation is prescribed. Meditation is the process of mind control and a variety of techniques for all levels of people are recommended by these religions. Self-realizatiog is the focus of life, meditation is considered more important than forcing others to believe in a particular ideology through force.
- There is sacred text known as a holy book for all the religions. In some religions, reading the sacred book on a daily basis is a must but in some, there is no such compulsion. But the holy book is

the repository from where one can know what good deeds are and what qualities must one inculcate to get closer to God and what should one do to get rid of the sins that they have committed.

- Most of the religioas have a messenger of God. Christians consider Jesus as Son of God, Muslims consider Muhammad as their last Prophet and Judaism considers Moses as the final Prophet.
- All religious practices are meant to remove our sin and to remove any barriers that ies between the devotee ahe God. All religions are trying to create inner peace and realization of self in various ways. Hatred causes anger. When one gets inner peace, hatred behavior leaves him.
- Majority of religions share the existence of miracles as the greatest proof of these religions. There are also many testimonies from devotees to substantiate this.

Similar to common traits, there are also vast differences among religions. In some religions, mere belief is enough whereas, in some religions, belief as well as how effectively they can bring people of other beliefs into their reign is also measured. Sub-sections of a single religion fight among themselves to show which religion is more powerful.

1.4 IMPORTANT FACTS ABOUT RELIGIONS

If you think that the existence of so many religions is a wonder, then you must know that there is more to it. There are so many astonishing facts about religions that will make us go insane. Either a religious follower or a non-religious person, everyone must be aware of all these facts that are listed below:

As per data statistics of the year 2015, one-third of the global population was Christians and they are the largest religious group. Islam became the next largest religion and Hinduism the third.

Islam religion has the youngest population and because of the high fertility rate, the Muslim population is projected to increase by 70% between 2015 and 2060.

The Christian population is also projected to increase by 34%. Forty million people are expected to switch into Christianity during the period 2010 to 2050.

Christians and Muslims are expected to increase their presence in sub-Saharan Africa. By the year 2050, four out of every 10 Christians in the world are likely to live in sub-Saharan Africa.

Current trends show that by the year 2050, there wilmore number of Muslims in the U.S than Jewish or Christians. It is also expected that India will have more Muslim population than Indonesia in 2050.

In Catholicism, which is a faith group under Christianity, there are patron saints for each and everything including beekeepers (Saint Ambrose), headaches (Saint Bibiana), and coffee (Saint Drogo).

The Catholic Christian population is high in Latin America. This country is home to more than 425 million Catholics. In other words, more than 40% of the world's total Catholic population is found in Latin America.

Women engage themselves with religion more than men. According to some estimates, 83.4% of women around the world identify with a faith group, compared with 79.9% of men.

Christians and Jewe the highly educated religious people compared to all other religions. Muslim women are lesser educated than Muslim men.

Any person can convert to Islam if they do these three things: declare a formal statement of faith in front of one to three witnesses; declaring that there is only one God and accepting Muhammad to be God's prophet; their conversion must be triggered from within and not by force.

Believers of some religion refuse to take medical treatment. For example, Jehovah's Witnesses aren't allowed to receive blood transfusions, and Christian Scientists often refuse all traditional medical treatment. They believe that faith in God will drive away all their pain.

Similarly, Mormons are not allowed to drink tea, coffee, or alcohol.

There are still some countries in this world that are controlled by a particular religion. The best example is the Vatican City.

Some sections of certain religions forbid using modern technology. For example, followers of the Amish religion do not use electricity or telephones.

REFERENCES

1. https://www.allaboutreligion.org/
2. https://simple.wikipedia.org/wiki/Religion
3. https://www.everystudent.com/features/connecting.html
4. http://www.religionfacts.com/
5. https://www.compassion.com/poverty/religion-quick-facts.htm
6. https://www.dosomething.org/facts/11-facts-about-world-religions
7. https://www.factmonster.com/dk/encyclopedia/society-and-beliefs/religion
8. https://www.theregister.co.uk/2006/10/06/the_odd_body_religion/
9. https://www.uri.org/kids/world-religions
10. https://www.bartleby.com/essay/Religion-Of-The-World-P3J4DC6ZTC
11. http://www.yourarticlelibrary.com/religion/essay-on-religion-meaning-nature-role-and-other-details-5931-words/8586
12. http://bigessaywriter.com/blog/three-major-world-religions-essay
13. https://www.collinsdictionary.com/dictionary/english/religion
14. http://atheisme.free.fr/Religion/What-is-religion-1.htm
15. http://agniveer.com/understanding-religions/
16. http://www.bbc.co.uk/religion/religions/
17. http://www.garvandwane.com/religion/religion1.html
18. https://www.ancient.eu/religion/
19. https://www.slideshare.net/desmondtwsa/buddhism-sociology

History Behind Religions

2

"All religion, my friend, is simply evolved out of fraud, fear, greed, imagination, and poetry."

–Edgar Allan Poe

CONTENTS

History of the religions takes us to the ancient space of humankind. Everything in this world has evolved from the time of creation and religion is no exception to that. There were different forms that religion took before coming into the form that we see currently. Many historical findings and cave arts give us a picture of how and in what way religions existed in the olden days. This Chapter will take us to the historical times when religion was created and how the stone age people termed it. We will also see how people of different civilizations and religions worshipped God and their belief systems.

2.1 BEGINNING OF RELIGION

Religion is known to exist from the upper Paleolithic age which existed about 50, 000 years ago. One of the most common traits for all the humans living across in this world is that everyone is involved or indulging in things which surround religion and its activities. We can easily presume that along with the other cultural aspects such as music, art, and communicative languages. But it is still unclear how religion set its origins.

2.1.1 Relationship between Agriculture and Religion

Researchers say that there could have been a lasting connection between agriculture and religion. It has been identified that before 10,000 years, the first ever farmers dwelled in the Mesopotamia. Mesopotamia is one of the historical regions in Western Asia whicdsright now corresponded te Southeastern Turkey, and regions along the Turkish–Syrian and Iran–Iraq borders. It could've risen because of the need for convenience or out of sheer necessity or out of the richness of the place they were settling down. Because up until then man was used to a nomadic way of living. This the researchers say could have been the building blocks of the religion.

Farming violates the way naturs,works; it is manipulating the environment to behave in a certain way to accommodate the human's way of living. Be it before 10,000 years ago or right now, even if the technological advancements of farming have helped the cultivation aspects to be better, farming cannot thrive without the universe's absolute necessity of sunshine and the rain. It is easy to decipher that there could be a connection between man, nature and God, for the mere reason that man needed the Sunshine and Rain and prayed to God to receive the natural blessings. It became an unwritten agreement between God, Man and Nature. As time went by, there was a specific day in a week or a month or a year where special prayer was made to honor God who helped them in their agriculture. Farming also paved the way for skilled people in praying who would be able to administer the contract at the special place where this was supposed to happen. One of the proverbs which correlaes this is “Pray Now, Eat Later “. Thre, therefore, explains the relationship between the systematic religion, the priests, tth Sabbath and the church.

2.1.2 What Cave Arts Tell us?

One of the interesting ways in which the beginning of religion in the Paleolithic age is the cave arts which are still existing in the Gargas Caves located in France. Some of the art in these caves bring light to the fact that the early cavemen had a belief of the fact that supernatural spirits and beings existed in these dark, deep and forbidden caves. Another interesting discovery of the Gargas caves is that there are around 230 handprints on the walls of these caves. The fact becomes even more fascinating when you realize that the 230 handprints incldes that of the adults and children.

Dr.Jean Clottes an internationally acclaimed expert on archaeological cave arts ma a mention that the handprints on these walls were basically negative hand stencils. Some of these prints had a missing finger which could be either because of a disease or the fact that it was broken during a hunt or sacrifice of a finger for their God. Dr. Clottes claims that these cavemen believed in supernatural powers and they used the caves for praying along with the rituals around it. These caves can be easily claimed as the churches, the first of its kind.

2.1.3 Origin

Religion is the set of beliefs created based he nature, the purpose for which the universe is in existence which also involves certain practices and rituals with a strict moral code by which the followers of that religion live by. Ancient North East is claimed to be the origin for the religion. The three major groups in which the religion can be categorized include polytheistic, pantheistic and monotheistic. One solo group which arose into existence after the “Enlightenment” period.

2.1.4 Polytheism

Polytheismis a worship or a belief of multiple God’s with an assembly of the pantheon of Gods and Goddesses. This includes the various rituals and beliefs associated with each of the God’s. Polytheism is told to be in present in this world since 2500 BC. Hinduism is one of the migionreligions where polytheism is practiced. Bhagavad Gita gives details of the Hindu beliefs which givesgive insight about how every other God mentioned were headed by the Brahma.

Polytheism also comprises of many ancient cultures from Assyria, Greece, Babylonia, Egypt, Rome etc. Per the Polytheistic beliefs the natural phenomenon such a rain, sunshine, fertility and harvests are in control by the many God's in the preamble. Another interesting aspect is that all the polytheistic beliefs had the practice of sacrificing for their God's to placate them in providing them what they require. Canaanites worshipped God's and Goddesses who oversaw each of the compartments of life and the planet. Baal the male God is responsible for the rain and harvest while the fertility and reprtion was were under the radar of the female God Ashteroth. Romans and the Greeks developed an even highly systemized pantheon of God's and Goddesses.

2.1.5 Pantheism

Pantheism is a doctrine whereby evthing in on this earth is a manifestation of God. Many of the ancient cultures showcased pantheism. The African and American Indian cultures considered universe was divine. This was engraved in the animism beliefs of these cultures. Pantheism also was the context of the old Egyptian times, then the pharaoh's d. In the some of the specific cultures of the far east Buddhism, Confucianism, and Taoism also based the beliefs and value system on the pantheism.

Some of the latest movements also try to bring back the principles of pantheism. The ultimate principle of pantheism is that God is everything and every little thing on this planet is of God's. Hence, there is a necessity for mankind to be in har with the nature, nurture it thereby to be nurtured by it in return. According to pantheism, humans are no different from animals and it is necessary for us to be at peace with them, take care of them, learn from them and develop the relationship between the humans and the elements of the earth.

2.1.6 Monotheism

Monotheism means that there is only one God who controls and directs the world. This is the underlying concept of Judaism, Christianity and Islam. All the three religion has its roots going to Abraham in 2000 BC. Apocrypha of the Jews, Koran of the Muslims and the Bible of Christians give ormation of about how God kept in touch with the people of the world. There is also evidence in these holy books as to how Abraham's descendants, the sraelites where led out of Egypt by Moses to their promised land, Canaan. For the next 1500 years of God's leading of the Israelites and the other surrounding nations which came in the way of Israel. This is elaborated in the old testament of the Bible. There are also specific laws and history of the Israel present in these scripturen the roman Roman empire was in their highest the Messiah who was prophesied was born in Bethlehem. In 32 AD Jesus completed his ministry and was crucified. On the thhe was risenrose from the dead and he ascended into heaven to be seated on the right side of God the father. This is recorded in the new testament of le.

MuhammedMuhammad then preached in places around Mecca after 600 years of Jesus'son. MuhammedMuhammad claimed and was regarded as the supreme prophet from Allah himself and his teachings became the major precepts for the Koran.

Important Dates in the Origin of Religions	
Year	**Historical Event and the leaders**
2000 BC	Time of Abraham, the patriarch of Israel.
1200 BC	Time of Moses, the Hebrew leader of the Exodus.
1100 - 500 BC	Hindus compile their holy texts, the Vedas.
563 - 483 BC	Time of Buddha, founder of Buddhism.
551 - 479 BC	Time of Confucius, founder of Confucianism.
200 BC	The Hindu book, Bhagavad Gita, is written.
2 to 4 BC - 32 AD	Time of Jesus Christ, the Messiah and founder of Christianity.
32 AD	The Crucifixion and Resurrection of Jesus Christ.
40 - 90 AD	The New Testament is written by the followers of Jesus Christ.
570 - 632 AD	Time of Muhammad, who records the Qur'an as the basis of Islam.

Figure 2.1: Important Dates in the Origin of Religions.

Source:https://www.slideshare.net/RobinHoltMclaren/world-religion-13017771

2.2 RELIGION IN THE ANCIENT WORLD

According to Marcus Tullius Cicero (106–43 BC), religion means to repeat or to repeat again or to show respect for what is sacred. He also emphasizes that beliefs or practices transcending spiritual experiences is what religion is all about. Therefore, one can easily deduce that every culture in human history has been practicing some form of religion. The modern age term "Mythology" in ancient times contained the facts of religion and rituals in an indistinguishable manner. These consisted of regular rituals and practices of the beliefs the ancient folks had on the supernatural beings who created the universe and all the other planetary objects in the cosmos.

Another interesting aspect which we understand from these mythologies is that all these supernatural entities had anthropomorphic aspects. Most of these aspects would correlate with the values of cultures of the adjacent geographies. For example, the ancient Egyptian religion had beliefs and practices ranging from magic, science, medicine, psychiatry, spiritualism, herbology and even concepts around life after death. Some of the ancient religions, however, like the rituals of the God's of Greece was totally opposite to the cultures of the land. Religion on the surrounds itself on the whole concept of the spirituality of the humans, the creation of the world and the creatures along with the incarnation of these God's. It also included the human beings place on the earth, the ways to overcome the sufferings in this world and the upcoming birth. Life after death ane eternity. Every nation's perception oe God's was different and the images or abstracts or idols looked different.

According to the Greek philosopher Xenophanes of Colophon who lived in the time period of 570- 478 BC opines that mortals conclude that all the God's and the Goddesses look and resemble themselves, he says that if the animals and the birds of this world had the capability to draw, write or make idols of God, they would do the same in the images of themselves, or an extended creative structure which looks like themselves. Xenophanes could have made this opinion given that the God's of the Ethiopians looked black and were flad,nosed; however, the Thracians' God's were blue-eyed and red-haired.

Xenophanes therefore believed and emphasized that there could only be one God who is the greatest ohethe entire God and Goddesses and the men of the world. He also claimed that God could look different. At the time his opinion was engulfed by other voices since he was a minority to raise the monotheistic view. People of that generation found it difficult to understand

Xenophanes and the only sect of people who got a hand os these thoughte was those thinkers and prophets of Judaism.

Most of the archaeological records gies a clear indication that our ancestors believed in many God and not just one. Every God had a unique and special sphere of influence. Therefore, it comes tn the assumption that every compartment of life requires help from different people to achieve a peaceful and wholesome living.

2.2.1 Perception of Ancient People

Every human being in this world must have an interaction between parents, siblings, teachers, friends, lovers, employers, doctors, gas station attendants, plumbers, politicians, and veterinarians. It is imminent that everybody must have interactions to live their life. And these interactions occur with the specialists of each arena not specific to one individual. This was the base of the ancient folks whereby they decided that the troubles in the different arena of their life needed different God's who can address their problems accordingly. The assumption is that since one won't take their dog to the plumber when it is sick, likewise God of war will not be able to give you advies of the issues around love and relationships. Hence, people with heartbreak went to the God of love and those who wanted to win in the war went to the God of War.

Figure 2.2: West Pediment of the Parthenon.

Source: https://www.ancient.eu/religion/

It was a common thought that every God who specialized in their areas made sure everything went well. Some of the cultures also shws how a certain God or Goddess would be so entwined with values of the people that they become so powerful and all-encompassing thereby converting the polytheistic nature into a henotheistic one. (As per Wikipedia," Henotheism is the worship of a single god while not denying the existence or possible existence of other deities".)

The difference between polytheism and henotheism is that in polytheism the masses worship many God's to fulfill the wide range of needs and wants and in henotheism one God is worshipped in different forms and incarnations. This rare and unique shift in the ancient world happened with goddess Isis and God Amun of Egypt. This is one of the striking examples where one of the mod'sGods's become the supreme and sovereign of the world in different forms.

Figure 2.3: Queen of the Night (or Burney's) Relief, Mesopotamia.

Source:https://www.ancient.eu/religion/

As mentioned in the earlier discussions it is evident that every culture had led their life abiding on a certain religion, however, to filter out the exact time when this developed is difficult. One of the mind-boggling puzzles is that if Mesopotamian religions had inspired the Egyptian religions or vice versa. We are no closer to solve this puzzle as every culture developed their own beliefs to explain how the natural phenomenon of the world works. This could've risen as a response to understand what happens in their life (example, how day and night occurs? or how the seasons change?) and sometimes the unfortunate event which unwhirled a series of unhappy events in humans' lives.

2.2.2 Religion in Ancient Mesopotamia

Mesopotamia is considered as the "Cradle of Civilization" with all the inventions and the cultural advancements which simultaneously is also the birthplace of religion. Sumer in the year 3500 BC has mentioned about religious practices in his written records. Per the ancient Mesopotamian religious practices, human beings were pars with the God's at work. It was the responsibility of the huans to labour and hold back thaos of the nature which was checked up by the supernatural power -God since the beginning of time. One of the historic myths says that the order of the world came into existence because of the pandemonamongst the God's. This is best demonstrated by one of the illustrations whereby the great God Marduk defeated Taimut and the forces of chaos to create te world.

The Ancient World: A Social and Cultural History by D. Brendan Nagle gives us more information about that even if the God's had defeated the forces of chaos and they had no power to bounce back and overrule the order of the world, it is impnt that the God's and the humans work in unison to restrain the power of chaos. Each of them had a unique role in this dramatic battle. However, the most important responsibility of the Mesopotamian dwellers was to provide everythieeded for the God's to run the world.

To have a mutually beneficial end in this world, every ancient Mesopotamian believed that they must worr and with the God's. Some of the historians had made a claim that these Mesopotamians were slaves to their God's. This claim itself seems absurd as every Mesopotamian thought that they are crkers with the God himself to help him take care of their daily needs. It is also recorded that their God's supplied them with beer which is the drink of God's. These God's were entwined with the lives of their people and lived in temples built by the people and not in some heavenly realms. Every Mesopity istherefore, therefore, an abode of temples. Most of these temples were structured in a way that it had towering ziggurat. Since the people considered this as the home of these God's each ofe statues were was bathed, clothed and fed by the priest or the priestesses. These statues itself were treated like Kings and Queens left in these temples like palaces all by themselves. As an illustration, we can consider Marduk statue being taken out once in a year oncasions and to honor him. Then the statue would be taken around the city of Bon so that the God can get some fresh air and to revel in the beauty of the city.

Inanna is another famous Goddess of this region who was revered for her specialty in love, sex, and war. She was another Goddess whose statue was greatly well-maintained by the priests and priestesses. She is the example of God's who go into the underworld and surfaces up to bring fertility and abundance to the crops. Across the southern regions of Sumer and Mesopotamia considered Inanna to be one of the most powerful and popular deity. Some of the other names by which she was called by different regions include Ishtar of the Akkadians (and later the Assyrians), Astarte of the Phoenicians, Sauska of the Hurrians-Hittites. She was also connected with the Aphrodite of the Greeks, Isis of the Egyptians, and Venus of the Romans.

Figure 2.4: Ishtar.

Source: https://www.ancient.eu/religion/

In the Mesopotamian history starting from the Akkadian Empire (c. 2334-2150 BCE) to the Assyrian (c. 1813-612 BCE), temples were a source of the city's life. These temples served as multiplexes where the priests disbursed the grains and excess goods to the poor, the counseling and medical centers, and the places of fun in times of festivals for their dedicated God's. Mesopians view of the life on earth looked cool as it was being nurtured by these Gods. However, the afterlife of these humans looked unsophisticated whereby the souls returned to the underworld or far away mountains drinking contaminated water and eating dust. These places were named "Land of No Return". This miserable picture of the underworld was totally different from those that of the Egyptians.

2.2.3 Religion in Egypt

The most important similarity of the Egyptians and the Mesopotamians in their way of worship considerihemselves as the co-workers with God's to maintain the balance in the Universe. Maat, which also known as the harmony principle is the one which bound the religion of Egyptians. They also made it a point to believe in the afterlife. Egyptian religions focused on and integrated every aspect of life. It is an amalgamation of magic, mythology, science, medicine, psychiatry, spiritualism, and herbology. The religion also has modern views of belief in a higher power to get things in place. God's according to the Egyptian mythology made sure that everythins provided to the human beings to have a good life. A perfect life in the world and in the afterlife when they die.

Figure 2.5: Chain of Ankhs.

Source: https://www.ancient.eu/religion/

The history of the Egyptian religion's dates to 3400 BCE, which is around the predynastic period in Egypt (c.6000-c.3150 BCE). The mostent forces of the God's who had at that time were popular as Isis, Osiris, Ptah, Hathor, Atum, Set, Nephthys, and Horus. The creation story of the Egyptian world is like the Mesopotamiarling chaotic waterswere only slow swirling chaotic waters. Out of the deep blue ocean a mound of earth came up which resembles the now pyramid, this primordial mound is also known as ben-bentum, the sun God in is known to stand on the ben-ben and spoke making the God of Shu (air), the Goddess Tefnut, the Goddess of Moisture, the God of Earth – Geb and the Goddess of the Sky – Nut. Heka which is known as magic accompanied Atum on the mound ben-ben to give birth to the whole of the universe using magic.

During the creation Atum decided that Nut, the Goddess of the Sky would be wife, however, Nut fell in love with Geb, the God of earth. Atum was pissed off and decided to separate them. He displaced Nut to the high end of the sky, so she wouldn't meet Geb. However, during the sunset, these two were able to meet and gave birth to three sons, Osiris, Set, and Horus, and two daughters, Isis and Nephthys. Osiris the eldest son was declared the Lord of the earth and Isis was wed to him. The second son Set was filled with jealousy and killed Osiris to claim the throne. Isis embalmed Osiris's body and ensured a powerful charm was set that Osiris came back to life. Osiris then became the supreme judge of the souls that are dead in the hall of truth to assess if their heart had the crediby to enter into the eternity.

Field of Reeds was the name given to the Egyptian afterlife and was mostly mirrored to the life which happens on the earth. From the favorite tree to the stream or the dog, everything was destined to follow the human at the time of his death. Egyptians foresaw the afterlife as part of the wholesome living of the humans. They made all efforts to ensure that the afterlife was smooth as well. This is very much evident from the tombs or the pyramids, temples, and funerary inscriptions such as the Pyramid Texts, Coffin Texts, and The Egyptian Book of the Dead. This, the ancient Egyptians thought it to be the soul's passage to the next world. Per the Egyptian myth, the Goddess Qebhet oversaw quenching the thirst of the souls in the underworld. Serket and Nephthys were the caretakers and protectors of those souls which journeyed through the Field of Reeds. Hence, it was a very well-known datum that from the birth of a human being until his death, then from his death to the course of his afterlife was systemized by the God's and run by God's. The principle of order is set in a unique way.

2.2.4 Religion in China & India

One of the oldest living religions in the world is Hinduism. The principle of order is highly dominant in the religions of these countries. Hinduism adheres to Sanatan Dharma, 'Eternal Order'. Hinduism is mostly considered to be a polytheistic religion. However, Hinduism is henotheistic. The only supreme God of Hinduism is Brahma and all the other God's are reflections and other aspects of the supreme God Brahma. Brahma is a very vast ocean and the human kind sometimes finds its incomprehensible. Hence, he found different forms to get in touch with people. Vishnu and Shiva are the most important representations of Brahma. The Bhagavad Gita alone showcases 330 million different deities known to be existing in the religion Hinduism. This includes God's such as Krishna who is well known throughout India and other lesser known God's who are speally revered in each geography.

Figure 2.6: Vishnu as Varaha, Udayagiri Caves.

Source: https://www.ancient.eu/religion/

Hinduism revolves around the concept that every individual has a specific order in the bigger order of the universe. Every human being on this earth is created to perform a duty or dharma which the individual only can perform. The righteous human beings of this world who can perform that duty assigned to them perfectly i.e. karma then they are made to move closer to the supreme God, Brahma. The final refuge would b become one with the God. If one is not able to fulfill his dharma in one lifee, then his souls is are reincarnated multiple times until the individual learns to live the right way and comes in union with the supreme soul.

Some of the aspects and beliefs of Hinduism was also part of the Gautama Buddha's teachings and life principles. He was also the founder of the religion Buddhism. According to the concepts of Buddhism, every human ing is looking for a closure, not with God but a higher nature. The major reason for this is that as per the teachings, every individual or a soul leaves behind an array of the suffering which clouds the mind with the fear of loss and death. One of the most famous religions of the older enturies which travelled to another country such as China is Buddhism.

4500 is the time when the ancient China developed a thought of religion. The Neolithic site of the Banpo village had some unique designs on the ceramics showcasing the time when religion could have been born in China. Animism and mythology could have been the two mixes of the belief structure because these relics have animals and dragons carved on it. As time went by and when the Xia Dynasty (2070-1600 BCE) came into existence, Shangti became the dominant God who was on the top of the charts with many anthropomorphic gods. Shang Dynasty (1600-1046 BCE) along with the existing religious practices also brought in to existence the worship of ancestors.

It was believed that Shangti was overwhelmed with a lot of responsibilities of the day to day life and sometimes was extremely busy to handle the needs of the people. Another unique aspect of these religions is the fact that the dead ancestors were the intermediaries between God and the living. Confucianism and Taoism were purely based on worship. Thistherefore, therefore, became the core tenets of these religions. As time went by Shangti was replaced by Tian, the heaven or the paradise where the dead would live forever.

Figure 2.7: Chinese Oracle Bone.

Source - https://www.ancient.eu/religion/

For an individual to enter into Tian or the paradise, the dead person must cross the bridge of forgetfulness over an abyss. One must look back at the life he has lived for one more time and drink from the cup. This will purge all the memories of the individual. At the bridge, based on the actions of the life, if the person deserved to get into the Tian, then he is moved on otherwise he is dragged into the abyss and gets swallowed. There are also versions of this fact where the soul is reincarnated after drinking from the cup. However, the death happened, the religion demanded its living beings to take good care of the ancestors.

2.2.5 Religion in Meso America

The belief system of the Maya also largely based itself on the remembrance of the dead. Ahe gods of the mayans Mayans were involvn every aspect of the human living. Like the other cultures, the God's of the Maya's had their own sphere of influence. They had about 250 deities for all aspects of their life. These God's wre said to be the one's who control the weather, the harvest of the people, the potential mates for the human beings and were even present at the birthdays of everybody and their death day as well.

REFERENCES

1. https://www.bartleby.com/essay/Religious-Beliefs-And-Practices-Common-Among-Indigenous-PK7M5AKVU5ZQ
2. https://www.britannica.com/topic/mystery-religion/Beliefs-and-practices

Nature- The Biggest of all Religions

3

"Look deep into nature, and then you will understand everything better."
–Albert Einstein

CONTENTS

'Religion' is something that most of the humans rely upon for inner satisfaction and happiness. Similar to religion there is yet another thing that we all depend upon for our existence and that is 'nature'. Even before the creation of man, nature existed. Nature is one of the most beautiful creations on Earth as well as the most powerful one. Before man knew about religion, he was living with nature. Hence it can be rightly said that nature is the biggest of all religions. This Chapter is intended to talk more about nature and how has nature merged in all our lives. This Chapter will also highlight the concept of worshipping nature as religion and the reason behind it.

3.1 NATURE, THE GREATEST TREASURE ON EARTH

There is nothing beautiful in this world than Mother Nature. God has created us and dropped us in the lap of mother nature like no other force in this world would be able to understand and satisfy human needs better than nature. Nature doesn't discriminate anyone; even a small ant with five senses is treated the same way as a human being having six senses. Nature nurtures every living being and non-living things equally. Nature is the one and only habitat for everyone.

Figure 3.1: Nature Gives Us Everything.

Source: https://www.pinterest.co.uk/pin/390265123938605505/

Whatever we eat, drink, breathe, everything is given by nature. We eat vegetables, fruits, leaves, we breathe oxygen, we need land to build houses for our shelter; from where do we get all these? Nature cares for us; it gives us everything in abundance. Ws, humans, consider the materialistic treasures such as gold, diamonds, pearls, emerald, ruby and many other valuables as priceless but all these gems are given to us by nature. If there is no nature, where would we find these valuable ? This itself proves that nature is the biggest treasure on Earth.

Nature helps everyone in different forms and it plays different roles. The fresh air inhaled in the wide open space will do wonders to a depressed soul. Naturs teaches us many things but it is we who fail to notice it amidst our busy mechanical life. There is no life on Earth without nature. It is the mother of all mothers. We are all blessed with nature. Waking up in the morning hearing the chirpings of birds and beautiful sunlight will give us instant energy than any health drink can provide for a full day activity. Why ds people who go walking and jogging in the morning seem energetic and fresh? It is because they start the day with nature. The scintillating sceneries maks our mind calm and peaceful.

Figure 3.2: Running During Early Sunrise.

Source: https://www.pinterest.com/made2travelblog/early-morning-run/

When we say that nature is the greatest treasure, we mean that it gives enormous benefits that are priceless. Still, some has made an attempt to calculate the monetary value of nature since people of today are so money minded, that only if the value of nature is said in terms of the money they would realize its value. BBC Earth study has calculated the economic value of animals, plants, and ecosystem keeping a price tag for everything.

The way the BBC Earth did it was pretty amazing. After realizing the Earth's index, the BBC network put forth nature's value in the stock exchange and published it in the financial sections of the Wall Street Journal (US), Times (London), Singapore Business Times (Singapore) and the Economic Times (India). This was the first time nature was listed in the stock exchange.

The index has been calculated in collaboration with the British environmental specialist Tony Juniper, commissioning a scoping study from the United Nations Environment Program World Conservation Monitoring Centre (UNEP-WCMC). "We scoured the available research and data to find comparable figures to company stock listings and settled on the financial value delivered each year," said Juniper. "What this unequivocally shows is the major contribution that nature makes to our health, wealth and security."

The results are given below and it is really surprising:

Figure 3.3: Planktons.

Source: https://www.lifegate.com/people/news/the-value-of-nature-earth-index

A group of different species of protozoa, algae, bacteria, and eukaryotic organisms which are called the Planktons is estimated to be of worth 222 billion dollars whiis are surprisingly more than the annual turnover of Apple, Microsoft, or Nestle.

We know that bees and other pollinaoris are very essential for agriculture and other activities. But it is really surprising that the services that they is are worth approximately 106 billion dollars. A single colony of them has an annual value of 141 dollars. We all know beavers but not its importance; One individual of this funny rodent, which we might think it could not influence our lives, annually produces services worth 120,000 dollars, for its contribution to reduce the risks of flooding and improve water conservation.

Figure 3.4: Bees.

Source: https://www.lifegate.com/people/news/the-value-of-nature-earth-index

We can hear numerous people seldom reiterate the importance of trees and forests. Trees help in inhaling Carbon dioxide and releasing Oxygen which the humans breathe. Such beneficial trees are valued more than $16 trillion. Roughly more than 1.6 billion people are depending on forests for their livelihood.

Figure 3.5: Trees in the Amazon Forest.

Source: http://www.sciencemag.org/news/2015/03/amazon-rainforest-ability-soak-carbon-dioxide-falling

According to the Earth Index, freshwater which is very essential for human survival is worth the total global economy. The total aggregate value of the water-related services provided by Nature is presented as at equivalent to global GDP, i.e. about $73.48 trillion.

Figure 3.6: Freshwater Flowing from the Hills.

Source: https://www.globalcitizen.org/en/content/where-in-the-world-is-all-the-fresh-water/

This study is conducted so that humans might understand the contributions of nature in a practical way. They would realize how much they are wasting and mistreating. S the nature is quite common to all, we do not care about it.

If nature was published as a listed company, we as brokers would have made a big issue of the senseless management of the natural resourcs.

EEach and every element found in nature is useful in some way or the other. There are five main elements of nature which is very important and which constitutes this universe. Those five elements of nature are – water, fire, air, space, and earth.

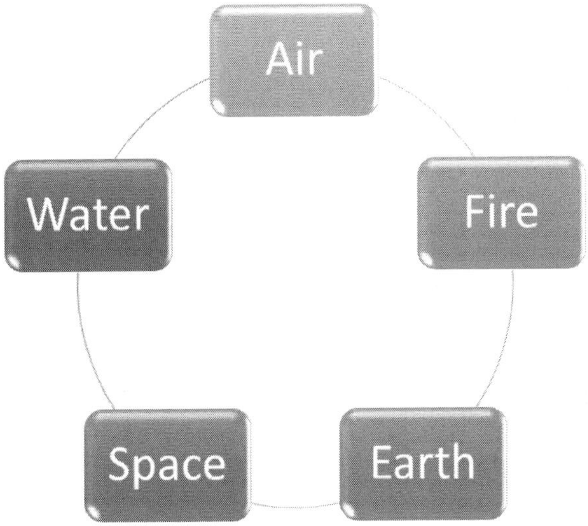

Figure 3.7: Five Elements of Nature.

Ancient scholars believe that the universe is made up of these five components of nature and these five components are equally balanced. As per Hinduism, nature, human body and religion are connected in a beautiful way and every person must know it. It is said that even the human body is composed of these five elements which we call as five senses of the human body. From a Hindu religious perspective, these five elements of nature are assigned separate Lords and Planets.

Figure 3.8: Balancing Five Elements in Your Daily Life.

Source: http://neetasinghal.com/balancing-the-five-elements-in-your-daily-life-2/

Let us understand the significance of these five elements of nature and how they are inter-linked with humans and religion. Learning this will help us understand the treasure that we have in nature and this will motivate us to preserve it.

Air

We know that our entire universe is surrounded by air and no living beings can survive without air. Air is always in constant motion and no one can arrest air. Air constitutes of various gases and elements. Oxygen is an important gas found in the atmosphere. Human beings need oxygen for breathing; the human brain needs oxygen to operate all the organs. Oxygen is also important to light fire. ***Lack of air means lack life in on the planet***. Trees are the main sources of oxygen. Nature has given us more trees which means more oxygen and we have to preserve those trees if we need to breathe.

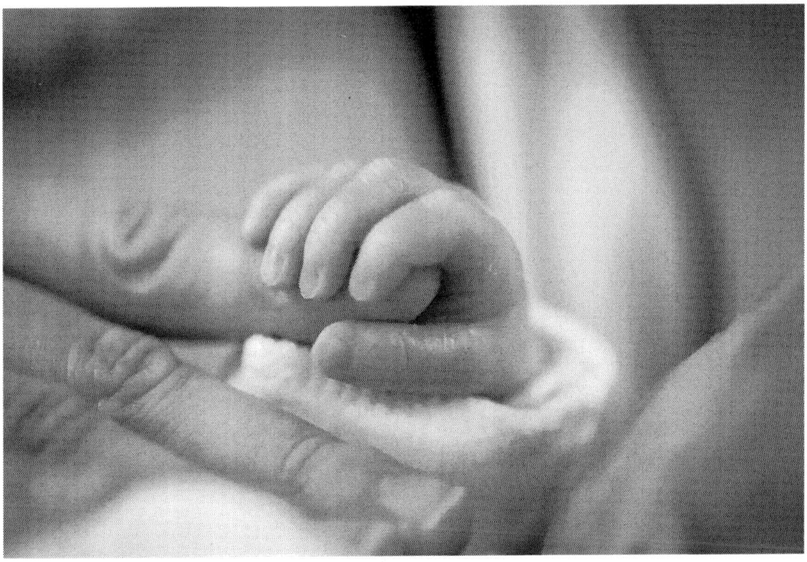

Figure 3.9: Sense of Touch.

Source: https://www.smallacorn.co.uk/fff/week-39/

Air is related to our human body. Air moves in our body as Oxygen. The presence of air is understood by the sense of touch or feel. The 'sense of touch' in our body is the skin and air forms the base of it. d is the a unique characteristic of air. The sense of touch and all other reflexive movements like twinkling and the contraction and relaxation of the eye are dominated by the quality of air. Everything remains perfect when there is a seamless flow of wind. Similarly, when the human body receives the adequate amount of oxygen it will function properly. There will be no laziness, ideas will start flowing, muscles will be strengthened and the body will get the right energy.

Water

Water is also one of the essential elements for survival. ***Water is one of the precious gifts of nature***. Plants and trees need water to grow. Animals and human beings need water to live in. Water is required for various other purposes. Humans generate electricity out of water. Water is used as a raw material for many industries and production houses. We cannot even imagine a life without water. Nature has given us so many sources of water such as lakes, rivers, seas and glaciers. Water that flows through mountains to the lands distributes minerals and make the land fertile. In the same way, the water running inside our body gives the body various nutrients, minerals and also carry away the wastes.

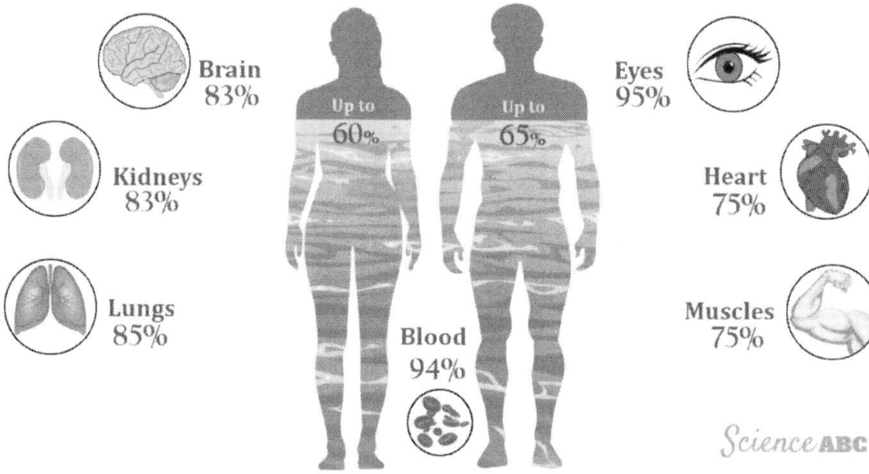

Figure 3.10: Amount of Water in Human Body.

Source: https://www.scienceabc.com/humans/long-can-survive-without-water.html

The way it is connected to our human body is interesting. Since water is in liquid form it is compared to 'juice' which signifies 'taste' and it denotes the tongue. The sense of taste and tongue are Water dominant in nature. The world is surrounded by water. Similarly, 70% of the human body coitutes of water. Water is very unstable by nature and it is in a liquid state as many other fluids in our body such as blood, urine, intracellular fluid, etc.,

Water is flexible in nature and can take any form or shape based on where it is poured. Similarly, human beings must learnhemselves to their surroundings. Adaptability and flexibility are very important to maintain healthy relationships with others. The feelings and emotions running in our also has have the qualities of water.

Fire

Fire is considered to be the source of allergies in on this planet. The stone age man invented fire and made him more civilized. ***Fire gives heat and light***. Light is very much required to drive away the ess. Fire The fire has the ability to melt and purify. It can transform solid to liquid and liquid to gas. The natural element of fire is 'Sun'. Without the sun, the entire Earth will

Figure 3.13: Space, An Element of Nature.

Source: https://theinspirationgrid.com/nightscapes-photography-by-matt-payne/

Space represents the infinite potential of the mind. It also denotes that a certain level of space is required in everything. Be it the space between parts of our human body or the space between relationships. For example, the space inside the intestines, tubes, arteries, etc., Even if there is a small variation in this space it may lead to adverse effects.

All the above analysis explains w beautifully nature is designed and how well it is connected to humans and it is also evident that no human brain could have done this. This shows the strength of nature. All these five elements of nature are very important for life as well as the human body. These five elements are also said to be the five states of energy. Balancing these elements is very critical. Even in our human body, when one of the five senses gets high or low, the othe senses works to balance it. We should also realize that all these important elements of nature when used in the appropriate way benefits the universe but when used without due respect, it has the ability to even destroy the universe.

3.2 WHAT DOES NATURE TEACH HUMANS?

We all go behind religion and are also very proud of being a follower of our own religions. Why do we go behind religion? It is because of the values and good morals that our religion teaches us. Nature too offers us many valuable lessons. Most times it is weumans who does not have the ability to untand what the nature is trying to say us. Every tiny creations of nature have huge vas to share to with humans. We do not blend with nature; we just use nature for our own requirements and move on. We do not even have the time to sit and watch what nature is and how all other creations of God are moving with nature.

Only if we intend to learn, we can derive the benefits out of nature. We can lead a happy and blissful life if we follow what nature teaches us. This section will explain evey moral values that nature and its cations teaches humans.

Being Humble

'Being Humble' is a great quality that every human being should possess in order to grow in their lives. Humility is considered by some as a passive quality, but it is not; one needs the courage to be humble. Nature teaches us to be humble. In fact, it gives us many examples to show us that it is not hard to be humble.

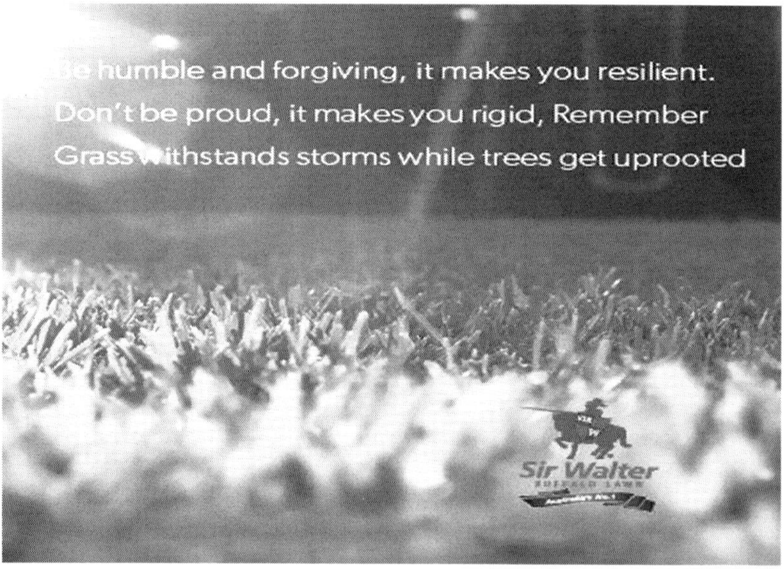

Figure 3.14: Be Humble as Grass.

Source: https://www.pinterest.com/pin/30540103693969598/

For example, the grass that we walk on t . The grass is so small and so many of them stamp on it every day. Still, it does not stop growing or stop giving comfort to people who sit on it. Humans must learn such qualities and be like grass. Nature is vast; what we see may not even be a quarter of it. There may be a lot of things that are unknown. In the same manner, human beings should accept the fact that there are many more things that are beyond human intellect and there will always be more new things to learn.

We should learn to keep ourselves grounded, respect and embrace our origins, and hopefully, give something back to those humble beginnings that nurtured us into who we are.

Everything Exists and Happens for a Reason

There is nothing called waste in this universe. Everything has a purpose behind it; there is some reason behind the existence of even the smallest particle. We can understand this concept easily if we notice our natural surroundings. If we take trees, for example, they provide shelter for birds, insects, and animals. It also provides fruits and acts as a source of food for humans and animals. When the tree dies, it is used to make furniture, it is also used as fuel for burning. The decomposed tree acts as a fertilizer for other plants.

Figure 3.15: Everything Happens for a Reason.

Source: https://www.pinterest.com/pin/549650329501209810/

Each stage of a tree has got a purpose. Similarly, every creation in this universe and every activity of them has got some purpose behind them.

They follow a beautiful cycle. Human beings always worry about every incident in their life. We have to look at nature and try to grasp the lesson that it teaches us. This is what even the religions preach – god has a beautiful reason behind his every act. One should just believe in Him.

Never Give Up

This is yet another importat mother natureMother Nature teaches us. Every day when we go out, we see many plants and flowers. We do not give much attention to them than just giving a glance at their beauty. But if we observe carefully, these flowers are not always present there. During the winter season, they wither and die; during spring they blossom again. Whatever the climate is they accept that change and still live again. There may be people who spit on the plants, pour dirty water on it, pluck its flowers. Does the plant stop growing or stop producing flowers? No, they stand up to all odds and still completes its purpose. They know they can live and so they do.

Figure 3.16: Nature Teaches Us Not to Give Up.

Source: https://www.pinterest.com/pin/531847037223518268/

Human beings must learn this attitude f nature. Now-a-adays, we see lots of people commit suicide unable to tolerate their failures. This not healthy to for the society. We should learn to pursue our interests, a career in spite of all the obstacles. We have to live with nature and imbibe these beautiful qualities in us.

Be strong

There are many things in nare that teaches us to be brave and strong. We should be strong and majestic as mountains.Mountains faces harsh winds, rains, different climatic disasters but it remains in its position unmoved. So should be the humans. Life m any situation,situation; we must embrace it and stand poised. Humans are emotional but we should not let our emotions rule us.

Figure 3.17: Nature Teaches us to Be Strong.

Source: http://andthenweallhadtea.blogspot.com/2011/01/amidst-skein.html

Always Hope for the Best

When we see seeds falling on the ground what do we think? We know that there is a maximum possibility for a plant to come out of it soon.

What do we think when we see the sunset? We know for sure that the sun is going to rise the next These are the valuable lessons that nature teaches us. It teaches us to hope for the best; it teaches us to think positively. The plant will sprout out of the seed when the process is complete, it cannot come out as per the plant's wish or our wish. Similarly, the sun will rise again based on the time cycle set by nature; the sun will not rise based on anyone's request. A fruit will fall from the tree only when it is ripe and ready to be consumed. What nature denotes is that each and everythi in life has a perfect timing and will occur accordingly.

Figure 3.18: Nature Teaches us to Hope for the Best.

Source: https://twitter.com/theboracaybeach/status/544300197266731008

Even the difnt seasons has have a natural flow. Spriollows the The bitter cold of winter that h and lirful changes ofand colorful autumn. Each of these seasons has its own importance and significance making the adage "All good things come with timeme with time.

Our duty is to do our work sincerely and hope that the best will happen. Nature trusts its creator and how can we humans not do it. Persistent behavior along with patience and trust will take us to places.

See Light in Darkness

The stars and the moon lit up the sky at night. They bring beauty even when it is dark everywhere. We should be like the stars and the moon. Even when we are surrounded by problems and sorrows, we should shine up and also make efforts bring light in to other people's life. When there are obstacles in the path of the river, it creates its own path and flows. We should also learn to create our own path to success and break out the barriers. Only when we experience darkness, we will be able to enjoy the essence of light. Sometimes, there is beauty in darkness also since it gives us the stars and the moon to overcome it.

Figure 3.19: Nature Teaches Us To See Light in Darkness.

Source: https://earthsky.org/tonight/seek-out-a-young-moon-after-sunset

Be True

There's more variety in nature. We learn different lessons from different elements of nature. The flock of birds that fly together teach us a very important lesson. What a beauty it is to see a flock of birds fly in the sky in a uniform way in the same pattern without running into each other? How is it possible for them at such an altitude? Do they have strict guidelines to follow when they travel? Who knows? But one thing which is very eminent is that they trust each other and they are in communication with each other even without really expressing it. This means they are able to understand each other so well and it shows how well they are connected.

Figure 3.20: Flocks of Birds Flying Uniformly in the Sky.

Source: https://www.nationalgeographic.com/science/phenomena/2014/01/15/birds-that-fly-in-a-v-formation-use-an-amazing-trick/

This quality is missing in humans these days. Actual bonding or true relationships must be like these birds. We must be able to communicate our thoughts even without uttering it. Partners should be able to get others sense of feeling from their body language and the tone of their voice even if they don't speak the truth. Thoughts are very powerful and it is easy to share our thoughts if we have a true bonding.

Treat Everyone Equally

Nature itself is beautiful; everything it owns is spectacular. Is it possible to say that a sparrow is more intelligent than a crow? Do they compare themselves like that? Even this thought seems funny and absurd. Nothing can be compared in nature. Everything in nature is valuable in a unique way and no one is better than the other. Humans must also treat everyone equally.

Figure 3.21: Treat Everyone Equally.

Source: https://www.brainyquote.com/topics/equality

When nature does not show any discrimination, why should we see differences among our own people and boost ourselves as the creation hing six senses. ? We should compare our success and growth only with ourselves and not with others as every human being is incomparably unique.

Unity is Strength

Bees and ants are the best examples to show the importance of unity and these two can be seen mostly near our houses. They work hard together and save for the future of their entire group. In order to save their life from the big fish, small fishes form a ball like structure. Most of us would have seen this in discovery channels. The small fishes are so clever; they know that if they are separate ey can become a prey very easily and when they are in a group, it is hard for the big fishes to get even one of them.

There are also many other animals that follow the same tactic to save themselves from big wild animals. Animals and even small insects know the importance of unity. Human beings even after seeing all these, fight among themselves and pledge their life to politicians. We should constantly repeat that everyone is important and has a certainsponsibility to for the effective functioning of this community. The strength of unity can easily break any big social barriers.

Figure 3.22: Small Fishes Safeguarding Themselves from the Predator.

Source: https://www.naturepl.com/stock-photo-blue-shark-prionace-glauca-feeding-on-anchovy-engraulis-encrasicolus-image01444991.html

Accept Change

If we look around us or even within us, we can sense that everything is changing. Butterflies are the best examples to denote that nothing is constant in this universe. The life-cycle of a butterfly visually proves that life is short and we ought to take different shapes and sizes during this short period and prepare ourselves to fly high with vibrant colors. The stage of a butterfly is more beautiful than that of the caterpillar.

Figure 3.23: Butterfly Embracing Change.

Source: https://authorbeckyjohnen.wordpress.com/tag/butterflies/

Snakes too shed their skin often to get better. In the same way, as humans grow they gain more wisdom. We should be ready to accept and appreciate change, only then we can gain more knowledge.

Allow the Clouds to Pass

The thoughts going in our mind is like the clouds in the sky. The clouds keep on moving and at varied speed. Likewise, our thoughts also have different frequency and speed depending upon our mood and surroundings. Sometimes, we get clouded with negative thoughts and concentrate more on them. Doing so will only deepen our thoughts and make us weaker. We should not allow that to happen. As the cloudin the sky passes, the bitter movements will also pass. Once the rain stops, we will be able to see the clear sky and the bright sun that is shining high. Birds are able to fly in the sky passing all the clouds even if they are rain clouds. They do not get scared. Even if the clouds try to overshadow the sun, the sun pushes itself out shining brightly. We should be like the birds and the sun overcoming the different clouds in our life.

Figure 3.24: There is Always Clear Sky Behind the Clouds.

Source: https://www.self.com/story/17-things-you-should-know-the-next-time-you-say-im-so-depressed

Be Selfless

Everything present in nature is beneficial to others. We saw the benefit of tree in the above sections – the benefit of trees can be longlisted but the tree does not take anything for itself. From its trunk to the fruits, it gives everything for the world and takes nothing in return. When birds or somanimals sense a danger, the first thing they do is to make noise and alert its gang not thinking much about their own life. They sacrifice their lives and save others.

Figure 3.25: Being Selfless.

Source: https://www.quora.com/How-does-nature-inspire-you

We should thank nature because it teaches us how to live selflessly. When we offer services or any help to others without any selfish motives, our life will become happy.

Being Real is Beautiful

We are surrounded by nature which consists of the flowers, trees, sky, clouds, different birds, animals, reptiles, insects, rain, snow, breeze, sea, river, ocean, waterfalls, and the list goes on. Nature does not need anything else since its real and so it looks magnificent. Nothing can match the beauty of nature. It is flawless in its own way. Human beings too do not require any fake behavior or artificial colors to show them as beautiful. It is enough if we remain as what we originally are.

Figure 3.26: Being Real.

Source: https://twitter.com/uppalawanna3/status/641967861208973313

Being natural is the most beautiful ornament that one can wear. Fakeness will fade off but the real self stays as long as we live. Instead of indulging in hings that makes us shine in front of others, we need to let ourselves be who we really are. This is what nature also teaches us.

Small Things Mean a Lot

There are many things in nature that are tiny, yet spectacular and remains a visual treat as well as delight us. Small mountain dew, a drop of snow that falls on us, a butterfly, a light breeze, a tiny feather, smile of an infant – all of these gives us immense happiness. What nature teaches us is that we should learn to appreciate small things in our life. Dreaming big is good but we should not ignore the real happiness in the process of achieving big.

Figure 3.27: Small Things Means a Lot.

Source: https://eng.papajaime.com/appreciate-small-things

Follow the Routine No Matter What

The Sun rises in the morning and sets in the evening. This is the routine process that the Sun follows according to the laws of nature and the process never gets delayed. All birds wake up so early in the morning no matter if it had slept properly at night or not. They follow punctuality and discipline and do not blame others. But what we humans are doing. We often blame others for our mistakes. We should learn to be punctual and disciplined from the Sun and the birds.

Figures 3.28: Seasons Follow the Same Routine.

Source: http://apps.usd.edu/esci/creation/age/content/natural_clocks/leaves.html

Live with Nature

Nature teaches us so many important lessons that one needs to follow to lead a blissful life. So, in order to learn all these lessons, we should be able to live blending with nature. Today the world has advanced and modernized in a way that we are moving away from nature. We disregard nature and the benefits that it provides. There are so many medicines available in nature that our ancestors used for all the diseases but we have turned deaf ear to all those and are going behind pills that cause harmful side effects. The more we go away from nature, the more we create trouble for ourselves. Nature has the potential to do anything. Nature can give us happiness, remove our depression, can give us hope, relief from ailments and what not. Nature has certain lthat are not man man-made and it acts according to these laws. It is up to us to worship nature and live with it or move away from it and face all the adverse consequences.

We should learn to embrace nature and listen to what it teaches us. There is no great teacher or religion than Mother Nature.

3.3 LINK BETWEEN NATURE AND RELIGION

Scientists are on continuous research to find out religion or the religious thought originated. There are many findings and theories on this topic.

Kelly James Clark, a Senior Research Fellow at the Kaufman Interfaith Institute at Grand Valley State University in Michigan talks about the theory of how humans interacted with their natural environment and this is the most widely discussed concept about the origin of religion. Kelly explains that ancient humans developed a mechanism called Hypersensitive Agency-Detecting Device (HADD) that lets humans perceive that many things have "agency," or the ability to act of their own accord. This was supposed to be the cause of how a human reacts upon a stimulus suchtling in the grass. (Lions act of their own accod. Better run.)

TThis HADD also planted the seeds of religious thought in the minds of humans. Apart from thinking that the rustling of grass may be due to the movement of a lion, they also started to think that there is some purpose behind the their own accord. For example, they thought raindrops were "acting for a purpose," Clark said that is every action was looked as having a purpose.

The theory of thinking about acting for a purpose is called by scientists as the 'Theory of Mind' (ToM). This ToM often comes up in the discussions on the origin of religion. ToM was all about anticipating what other beings' actions might be and planning their own actions accordingly. When human beings started to turn their anticipation towards actions of non-actors such as raindrops, thunder and natural things like that, that is where the seed of supernatural thoughts started sprouting in the minds of humans.

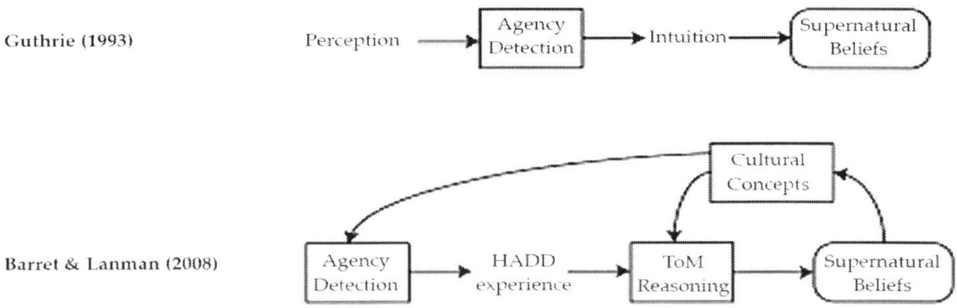

Figure 3.29: How HADD & ToM Works.

Source: https://philarchive.org/archive/VANSTS-6v1

Allen Kerkeslager, an Associate Professor in the Department of Theology and Religious Studies at Saint Joseph's University in Philadelphia states that,

"The roaring threat of a thunderstorm or the devastation of a flood is widely seen across cultures as the product of a dangerous personal agent in the sky or river, respectively. Likewise, the movements of the sun, moon and stars are widely explained as the movements of personal agents with extraordinary powers."

Thus, the ancient men started believing in the existence of supernatural powers. According to many scientists, ancestral spirits, goblins, fairies and gods formed the basis of relibeliefs and t kind of believes beliefs still exist. Kelly James Clark says that we all have a hyperactive agency-detecting device. When a man starts to explore about himself that is where religion begins. When he starts to identify his source, he starts looking at nature the source of which is also unidentifiable. Such a thought process can allow for a good associat in which he lived.

Ancient scholars believe that human beings have five layers of experience. The experience that they have with the environment, with their own physical body, their mind, their intuition and their inner-self or the spirit. Here the environment is nothing but nature and if our environment is neat and clean, it gives a positive vibe to our body and mind. The main purpose of religion is to calm our body and mind and to help us focus. This ultimately proves that both nature and religion are doing the same thing and that's why nature is always linked with religion and vice versa.

In all our ancient cultures, natural elements such as sun, moon, rivers, mountains, trees, etc., were given portance and mostly worshipped. That is why people who give importance to religion and nature, preserve their environment and when people start moving away from nature, they destroy it. Ancient people gave importance to both religion and nature and both were the same for them. They did not wash their clothes in holy water. Hindus used to dissolve the ashes from cremation grounds in the holy water because they believed that thing must be send sent back to nature. People living in ancient times never used to cut trees for unnecessary reasons; even if they cut one tree, they planted five instead. All the ancient traditions connected to nature with religion. But today, people are ignoring their traditions an so nature also. IIn the greediness to obtain more and succeed, people are not thinking about the harm that they are causing to the environment. This sort of behavior has led to violence that causes wars and concomitant long lasting damage testore and repair. Only when humans start to care for one another, they will care for nature. Caring for one another is what every religion teaches. There are many opposing things in nature – for example, water puts off the fire. There are also many animals that prey on other animals. Still, nature balances everything perfectly. Religion teaches this important quality of nature i.e., to balance the opposing forces within ourselves and outside us.

3.3.1 Importance Given by Religions to Nature

All religions talk about the creation of the universe and most religions believe that the creator of the universe is God. Hence all religions emphasize the importance of preserving nature which is the God's creationligionsof the world hasreligions of the world have to tell about nature:

Buddhism

Buddha's lessons insist that it is the responsibility of each one of us to conserve for the future and future generation. Many of the teachings in ddhist literature refersrefer to nature.

As a bee – without harming the blossom, its color, its fragrance – takes its nectar and flies away: so should the sage go through a village. (Dhammapada IV, Pupphavagga: Blossoms, 49)

Drop by drop is the water pot filled. Likewise, the wise man, gathering it little by little, fills himself with good. (Dhammapada IX, Papavagga: Evil, 122)

Our ancestors viewed the earth as rich and bountiful, which it is. Many people in the past also saw nature as inexhaustibly sustainable, which we now know is the case only if we care for it. (Dalai Lama, 1990)

Figure 3.30: Buddhism's Teachings on Nature.

HINDUISM

Hinduism is one religion that believes strongly in nature. If we refer to any sacred texts of Hinduism, we can find so many references about nature and that makes the followers believe that nature must be treated as part of our life.

"I shall now explain the knowable, knowing which you will taste the eternal. Brahman, the spirit, beginning-less and subordinate to Me, lies beyond the cause and effect of this material world. (Bhagavad Gita)

"According to the different modes of material nature — the mode of goodness, the mode of passion and the mode of darkness — there are different living creatures, who are known as demigods, human beings and hellish living entities. O King, even a particular mode of nature, being mixed with the other two, is divided into three, and thus each kind of living creature is influenced by the other modes and acquires its habits also." (Bhagavad Gita)

Figure 3.31: Hinduism's Take on Nature.

As per Hinduism, in order to protect the human race and all other living beings, one must protect nature. There is no life without nature.

Christianity

Christianity also believes in the concept of environmental protection like other religions. Christians have environmental responsibility which is preached by their religion. Bible has many verses stressing this fact.

Do not pollute the land where you are. Bloodshed pollutes the land, and atonement cannot be made for the land on which blood has been shed, except by the blood of the one who shed it." (Bible)

"When they had all had enough to eat, he said to his disciples, 'Gather the pieces that are left over. Let nothing be wasted." (Bible)

Figure 3.32: Christianity's Take on Nature.

Islam

Islam believes that the Earth is the creation of God and human beings are entrusted with td) of the God's creation. Similar to the Bible and Bhagavad Gita, the Quran (holy book of Islam) also has many verses supporting the protection of the environment. Islam religion prohibits its followers from the over-consumption of natural resources. In fact, wasting natural resources is said to be the thirty-second greatest sin in the Quran.

"Devote thyself single-mindedly to the Faith, and thus follow the nature designed by Allah, the nature according to which He has fashioned mankind. There is no altering the creation of Allah." (Quran)

"It is Allah who made for you the earth a place of settlement and the sky a ceiling and formed you and perfected your forms and provided you with good things. That is Allah, your Lord; then blessed is Allah, Lord of the worlds." (Quran)

"Do not strut arrogantly on the earth. You will never split the earth apart nor will you ever rival the mountains' stature" (Quran)

Figure 3.33: What Islam Says About Nature?

It is to be noted that in 2015, the Islamic Climate Change Symposium adopted the Islamic Declaration on Global Climate Change.

Judaism

According to the tradition followed in Judaism, the land and environment are properties of God, and it is the duty of humankind to take care of it. It is said in the book of Genesis that God chose Eden garden initially for the humans to live.

"And God said: 'Behold, I have given you every herb yielding seed, which is upon the face of all the earth, and every tree, in which is the fruit of a tree yielding seed--to you it shall be for food." (Genesis)

"The Earth is the Lord's and the fullness thereof". (Psalm)

"[...] the Earth is Mine, you are My tenants". (Leviticus)

Figure 3.34: Judaism's Take on Nature.

Jainism

Jainism is a religion with its root in India and this religion focuses on ahimsa i.e., non-violence. This religion teaches its followers to be disciplined, peaceful, kind to all living beings (that includes even insects and other animals), follow vegetarianism and avoid wasting. In 1990, The Jain Declaration on Nature was written to mark the entry of the Jain faith into the WWF Network on Conservation and Religion thus proving to the world the importance the religion gives to nature.

"Do not injure, abuse, oppress, enslave, insult, torment, torture, or kill any creature or living being." (Mahavira)

"As a highly evolved form of life, human beings have a great moral responsibility in their mutual dealings and in their relationship with the rest of the universe. It is this conception of life and its eternal coherence, in which human beings have an inescapable ethical responsibility, that made the Jain tradition a cradle for the creed of environmental protection and harmony." (The Jain Declaration on Nature, 1990).

Figure 3.35: What Jainism Teaches About Nature.

Sikhism

Sikhism is another Indian religion founded by Guru Nanak Dev Ji. The holy site of the Sikh is managed by the Shiromani Gurdwara Parbandhak Committrganization makes decisions for the global Sikh community, and they focus more on the environment. The sacred text of Sikhism known as Guru Granth Sahib also has many teachings related to the environment.

> **"You, Yourself created the Universe, and You are pleased…You, Yourself the bumblebee, flower, fruit and the tree." (Guru Granth Sahib)**
>
> **"You, Yourself the water, desert, ocean and the pond. You, Yourself are the big fish, tortoise and the Cause of causes." (Guru Granth Sahib)**

Figure 3.36: Sikhism's Teachings on Nature.

Shinto

Shinto religion is very close to nature and this nature is based on Kamis, spirits corresponding to natural entities such as wind, rocks, water, etc., The followers of Shinto religion are expected to live in union with nature as well as other human beings.

> **"I will give over to my child the rice-ears of the sacred garden, of which I partake in the Plain of High Heaven." ((Nihongi II)**
>
> **"The plan is for Religious Forests to be managed in ways which are religiously compatible, environmentally appropriate, socially beneficial and economically viable." (Jinja Honcho)**

Figure 3.37: What Does Shinto Religion Teach About Nature?

Taoism

Taoismlso is known as Daoism is an the oldest Chinese religion that believes in the harmony between nature and humani. The Dao principles teaches one how to behave and also lead others in the right way. The sacred text of Taoism talks about how nature teaches important values to humans.

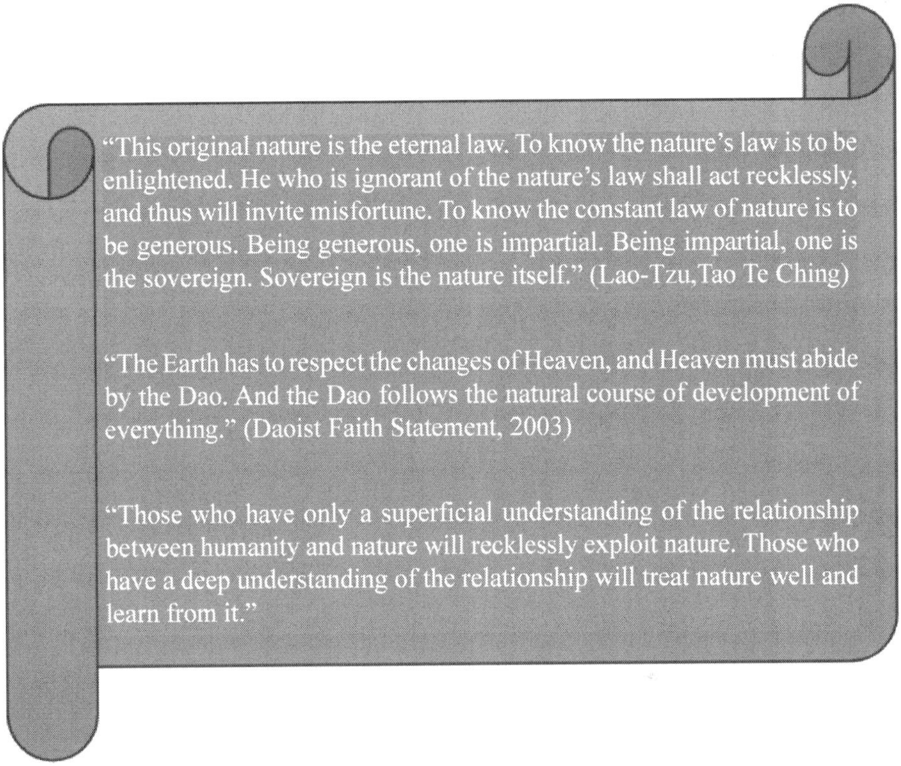

Figure 3.38: Taoism's Teachings on Nature.

Confucianism

Confucianism does not stop with influencing the culture and politics of China but it also impacted the people of Japan, Vietnam and Korea. Some sociologists referred to Confucianism as a civil religion or diffused religion. The Chinese considered Confucianism as their way of life. The principles of Confucianism also talk about the protection of nature and ecology.

> "... sustainable harmonious relationship between the human species and nature is not merely an abstract ideal, but a concrete guide for practical living." (International Confucian Ecological Alliance, 2015)

Figure 3.39: Confucianism's Take on Nature.

Baha'i Faith

Baha'i Faith works on preserving the environment so that the environment will be preserved and humans can live happily with nature.

> **"Nature is God's Will and is its expression in and through the contingent world." (Tablets of Bahá'u'lláh)**
>
> **"By nature is meant those inherent properties and necessary relations derived from the realities of things. And these realities of things, though in the utmost diversity, are yet intimately connected one with the other." (Bahá'u'lláh and 'Abdu'l-Bahá, The Bahá'í Revelation)**
>
> **"...to man God has given such wonderful power that he can guide, control and overcome nature.... What ignorance and stupidity it is to worship and adore nature, when God in His goodness has made us masters thereof." ('Abdu'l-Bahá, Paris Talks)**

Figure 3.40: Baha'i Religion's Teachings on Nature.

3.4 NATURE WORSHIP

From the above sections, it is clear that from time immemorial people have given importance to nature and have preserved them. Nature has been the cause of the evolution of religion andso most of the religions insists on the conservation of nature. In order to make people understand the importance of nature, most religions started worshipping nature. Sun, moon, water, and air were considered sacred and people worshipped them.

The belief system in which God is seen as the forces of the Universe is called Pantheism. But worshipping nature in this sense has not been documented anywhere in history. Nature worship in some cultuthat there is a high God who is theor Lord in Heaven (the deus otiosus or "hidden, or idle, god" in Latin) who has been withdrawn from the immediate details of the governing of the world gated all work on earth to what arnifications of the forces of nature. In the Dyola of e Dyola of Guinea. In such reliied natural powers, or nature spirits.

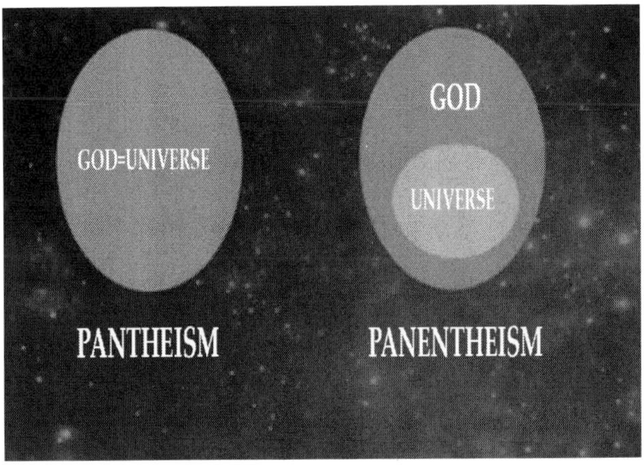

Figure 3.41: Pantheism.

Source: https://linkedinspirit.net/glossary/pantheism/

The worship of nature proves to the world that sacred can be in any form and holy things can be worshipped despite their form. Worshipping nature also reveals the connection that human beings have with nature and how human beings are able to identify the sacred that exists in nature. The concept of shared destiny of nature and humanity brings in the existence of emotions or pers, personalities, volition, and so on.

Differeligions have different believes upon in the elements of nature. Let us see some of them in detail:

The Sky

Most of the religions believe that sky is the dwelling place of Gods. People often pray to the sky god when they are in need of rain. Many communities believe in the sky god. A sect of people in East-Central Africa called the Konde, worshipped Mbamba (also callede Kiara or Kyala) as a divine force whothat lived with his family above the sky. It is the general belief that the divine forces of the sky initially lived on the Earth with other human beings and then moved to the sky.

Figure 3.42: Picture Representing Sky God.

Source: https://thequeenofheaven.wordpress.com/?blogsub=confirming

The people of Samoyed worshipped 'Num', a God said to have lived in the seventh heaven and whose name meant sky. The Samoyed people believe that the Num god covers the entire universe including the sea and the Earth. As we can see, the sky is called by different names by different people. Mongols called it the Tengri and the Buriats called it the Tengeri. The ancient tribes of South-eastern Australia namely the Euahlayi, Wiradjuri and Kamilaroi considered Baiame as their Supreme God and he is said to reside beside the flowing waters of the Milky Way. Thunder is considered to be the voice of Baiame and he took the soulsead people to his place. Such Althouand although they play a mjin initiatioies but , they do not usually dominate liturgical life.

Since the sky is considered sacred, the objects that fall from the sky are also meant to be sacred and these objects often become the symbol of religious cults. For example, the small pebbles of the Niger River Valley in West Africa called the Numana are considered to have been fallen from the sky. The people of these valley offer sacrifices to these pebbles and these pebbles are thought to be the fragments of sky god. Similar to the pebbles, meteorites, thunder stones and rain stones falling from the sky are also treated as sacred particles since they are believed to be the arrow points shot by the god of lightning or by other celestial divinities.

The Sun

Unlike all other Celestial things, the Sun is worshipped in most cultures and it is considered to be heroic. Everyone offer sacrifices to the Sun God. The Chukchi of Northern Asia, the Chagga of Mount Kilimanjaro in Tanzania offer sacrifices to Sun in times of crisis. People who rely on agriculture as their main occupation have great honor for Sun god as the Sun is connected with the fertility of crops. This applies even to the Inti in the Incaheon. For them, it does not limit with the fertility of food and crops but extends to human progeny. People who showed themselves as higher class people revealed themselves as descendants of the Sun God. The Inca nobles, the Egyptian pharaoh and the chief families on the island of Timor –all reckoned themselves as the children of the Sun.

Figure 3.43: Picture Depicting Sun Worship inAncient Egypt.

Source: https: //gaiasgold.com/sungazing-higher-consciousness-health/egypt-sun-worship/

According to the Apinage people of South America, it is the Sun and the Moon tngs fromof the Desana, a Tucano-speaking group of a southern Colombian a Tucano-speaking group the creation was via insminatione and caused the creation of the universe. In some cultures, people believe that the sun travels through the underworld at night. Therefore, Sun is seen as a sacred guide for the soul's journey through the land of the dead. Not only does the Sun transport the souls of dead waors but the dead cey Islands, the dead cluster in groups to await the biannual p trek or undeen the warriors who have fallen in battle.

The Moon

The Moon has been worshipped in many religions for a long period. Since the shape of the moon changes according to a particular cycle and its changing disposition in the sky makes it worshipped by all. The God of the Moon in Babylonia is called the Sin and it is said to have great connections with the waters of the Earth. The flow of the water was linked to the rhythmic capacities and periodic nature of Sin. It also created the grasses of the world.

Figure 3.44: Picture Depicting Moon Worship in North Babylonia.

Source: https://commons.wikimedia.org/wiki/File:Khashkhamer_seal_moon_worship.jpg

Nature worship is considered to be the oldest form of religion. From the above elaborations it is clear that apart from worshipping the natural elements such as wind, trees, water, mountains, animals, etc., humans also worshipped celestial bodies such as the stars, moon, the sun, planets, constellations, etc., which was termed as "astrotheology". And it is from this evolved astronomy and agriculture which helped in the development of human civilization.

REFERENCES

1. https://www.livescience.com/52364-origins-supernatural-relgious-beliefs.html
2. https://www.britannica.com/topic/nature-worship
3. http://rajhansnature.blogspot.com/2013/02/nature-god-gift-for-us.html
4. https://www.speakingtree.in/blog/nature---gods-most-precious-gift-to-mankind
5. http://www.saching.com/Article/Nature-is-beautiful-and-God-s-gift/2350
6. http://scholarlyanswers.blogspot.com/2013/06/nature-nature-gift-of-god-nature.html
7. http://www.traveling9to5.com/2011/11/nature-is-gods-gift/
8. https://www.lifegate.com/people/news/the-value-of-nature-earth-index
9. https://blogs.siliconindia.com/MissionMan/Water_Precious__free_Gift_of_Nature-bid-zVJPWi6835444444.html
10. https://brainly.in/question/2034495
11. https://astrobix.com/astrosight/704-significance-of-five-elements-of-nature-five-elements-of-nature.html
12. https://detechter.com/the-five-elements-of-nature-and-reasons-to-balance-them/
13. https://www.acharyashunya.com/single-post/2017/04/07/5-Great-Elements-of-Nature-That-Mirror-Our-Being
14. https://www.jewelsofhappiness.com/5-life-lessons-we-can-learn-from-nature/
15. http://wonderfulwild.com/12-lessons-i-learnt-from-nature/
16. https://www.purposefairy.com/68155/26-powerful-lessons-to-learn-from-nature/
17. https://missionself.com/important-lessons-we-can-learn-from-nature
18. https://www.togethercaringsharing.com/what-does-nature-teaches-us/
19. https://daveursillo.com/four-life-lessons-taught-by-nature/
20. https://www.huffingtonpost.com/sri-sri-ravi-shankar/our-spiritual-connection_b_648379.html

21. https://www.unenvironment.org/about-un-environment/faith-earth-initiative/why-faith-and-environment-matters/how-religions-are
22. https://www.encyclopedia.com/environment/encyclopedias-almanacs-transcripts-and-maps/nature-worship-nature
23. https://freethoughtnation.com/worlds-oldest-religion-honors-mother-goddess/
24. https://en.wikipedia.org/wiki/Earth_religion

Relationship between Religion and Environment

4

"I can find God in nature, in animals, in birds and the environment."
–Pat Buckley

CONTENTS

Religion and Environment are both associated with each other. It is said that one who respects his religion respects the environment as well. Both teaes the same thing. We all are surrounded by what we call the environment. Taking care of our environment is as important as praying to the divine force. If we ignore our environment, even the divine force will ignore us and we will be pushed to face the consequences on our own. In this Chapter, we are going to learn all about the environment and in what way is religion and environment connected. We will also discust the different ways by which religions help in preserving our environment and benefit the ecosystem.

4.1 ENVIRONMENT – DEFINITION AND INTRODUCTION

The environment is a wide term used to refer to everything that is around us. It need not be a living thing; even non-living things live in our environment and they all create a perfect balance. The term 'environment' is widely used by many and so there are many definitions and interpretations to it. In layman language, **environment simply means 'nature'**. For some, the environment means everything associated with nature i.e., the landscapes, mountains, wildlife and everything. In general, environment refers to the materials and forcthearound a liviismbeing.

Figure 4.1: Picture Representing Environment.

Source: https://www.game-debate.com/games/index.php?g_id=27578&game=Eco%20-%20Global%20Survival%20Game

The word 'Environment' is derived from the French word 'Environia' which means 'to surround'. So, the environment means the surroundings in which the organisms live which may be abiotic or biotic. The air (atmosphere), water (hydrosphere) and land (lithosphere) are all part of the abiotic environment. The biotic environment includes the plants, animals and the microbes. It is the environment that regulathe lsms. The environment has the perfect conditions to influence the growth and development of living forms. It not only surrounds but also impacts the livves. It is made of water, soil, air, sun ar, organisms that give us the requisite resourrgy. It has provided us all the resources for leading a comfortable lifnt The environment can be regarded a space or field where all of thestsexist.

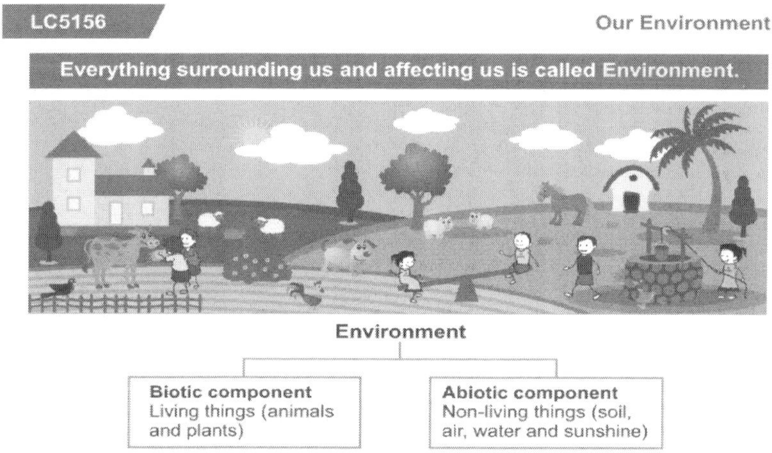

Figure 4.2: Classification of Environment.

Source: http://www.learnhive.net/learn/cbse-grade-6/science/our-environment

Man and environment are very close to each other. It is the human beings who depend more upon the environment and it can be said that more than any other living being, humans have more interactions with the environment. The environment existed when humans were born and so it is expected thhe humans should live in harmony with nature. The environment takes care of even the smallest needs of human beings. We depend upon the environment for Oxygen, food, shelter, medicines, etc., So, it is absolutely essential to maintain a healthy environment.

The environment is a reservoir of natural resources and conserving them is very important. The conservation of the environment means the conservation of natural resources. But what exactly happens is that humans try to modify the environment as per their requirements and try to control it. Whatever man does is only temporary; the changes that man does leads to potential loss of the environment.

Everyone has a different image in their mind for the environment and different beliefs associated with it. Though the definition and understanding of environment dffers from person to person, there is a central assumption as well as acceptance that there is some kind of relationship between environment and man as well as environment and religion. Hence the environment can be said as the backdrop to the unfolding narrative of human history, the habitats, and resources that humans exploit, the hinterland that surrounds human settlements, or the wilderness that humans have not yet domesticated or dominated.

There are many definitions to explain the term 'environment'. Let us see some of them in detail:

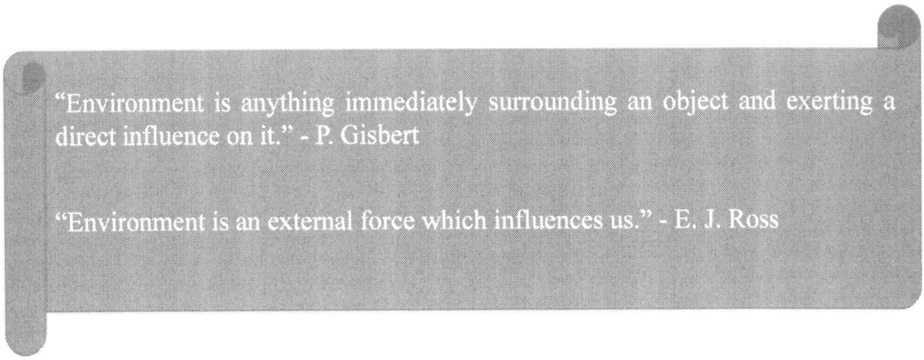

Figure 4.3: Definitions of Environment.

Environmental factors may be natural, artificial, biological, psychological and social. Although these may be distant from us, it has an influence on our lives. The environment affects a person's behavior, body andonment The environment is said to be everything that surrounds us at any given point of time. The reference to the environment changes depending upon the field of study. For example, Electromagnetic radiation refers to the radio waves and other lectro-magnetic radiation and magnetic fields. The galactic environment refers to the conditions between the stars. The environment can also be the biological environment, the physical environment, and chemical environment.

"The term 'elogical termand 'ecosystem', are used inte defined aswith the latter as a community of interacting organisms together with their physical surrou The notion concept of interrelationship is vital in the study of environment as one system influences in another.

The Environment can roughly be classified into two types such as Micro Environment and Macro Environment.

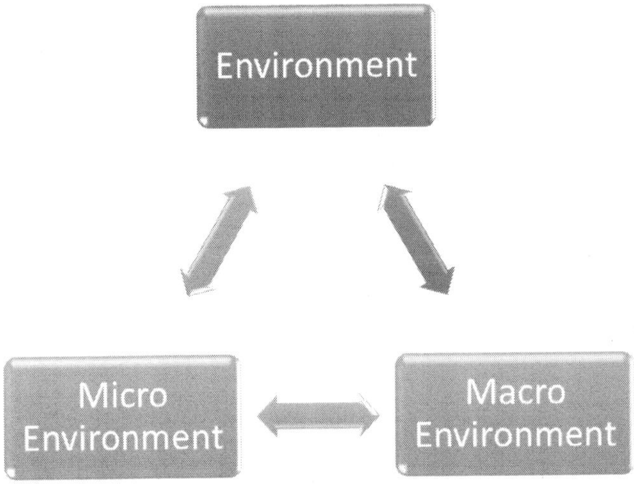

Figure 4.4: Types of Environment

Micro Environment

Micro Envirent refers to is the immediate local surrounding of the organism.

Macro Environment

Macro Environment refers to all the external physical and bihat surround thearganismexternally. Physical environment refers to the abiotic factors such as rainfall, light, soil, minerals etc., and the biotic conditions refers to the biotic factors such as plants and animals.

Today, people are focusing more on the environment. We can see discussions and debates on the preservation of the environment in televisions, newspapers nd social medias. Nations conduct Global Summits on a regular bs to discuss on prevailing environmental issues. More awareness programs and campaigns are conducted to make people understand how we are connected to the environment and the need to preserve them. These show the growing significa among people on in the environment and how our lives are tied to our environment.

Scientists are constantly examining the ways by which humans are causing harm to the environment. Air pollution, water pollution, acid rain, deforestation and many other problems that cause dangerous problems to both humans and the environment are caused by humans. So, when people talk about the environment today they refer to the overall condition of the planet.

4.2 LINK BETWEEN RELIGION AND ENVIRONMENT

Religion and Environment are closely relatother. Environment The environment here can be referred to as 'nature' and nature has always influenced religions and religions has always paid respect to tronment and learnt learned a lot from ill. We have learnt learned from the previous chapters that the concept of religion has rooted out from nature. Religion is also stated as the relationship between people and nature. Most of the people across the world follow a particular religion. There are approximately 750 million Hindus, 1.4 billion Muslims, nearly 2 billion Christians, 10 million Jains, 700 million Buddhists, 12.5 million Jews, 16 million Sikhs and 5 million Baha'is and all these people share a common practice i.e., to live in harmony with nature.

Figure 4.5: Different Religions and Environment.

Source: https://www.eoi.es/blogs/carloscerdan/2012/01/08/dp-4-relationship-between-religion-and-environmental-conservation/

Religion forms the base for everything that a human does. Right from the way we live - to the food we eat, our legal systems, national policies and constitutions – there is the influence of religion in everything. Environment plays a similar role in everyone's life as that of religion. Religion and enironment thus forms the basis of human development. And surprisingly both are seen sacred and has a spiritual connection.

Figure 4.6: Religion and Environment Teaches Everything.

Source: https://www.thestar.com.my/metro/community/2015/04/23/taking-care-of-mother-earth-firm-comes-forward-to-help-students-in-gardening-activity/

The Baha'i faith preaches that the diversity oe natural world are is the reflections of God. There is generally this belief among people of religions that the nature which is our environment is created by God, so it is sacred and it needs to be protected. In animistic religions, people believed that a spirit lives in every tree, mountains and hills and so everything must be respected. According to older religious traditions, even humans form a part of nature.

Figure 4.7: The Sacred Forest of Quidah, a High Spiritual Place.
Source: https://www.flickr.com/photos/lindadevolder/32692319910

The importance given to religion and environment is increasing day by day. Researchers and scientists who are continuously monitoring the environmental crisis are of the opinion that most of the environmental problems are caused by humans. With this in mind, they look to religion for rescue assome individuals holds a large degree of responsibility only towards religion and thus can work on the environmental problems. The essay, "The Historical Roots of our Ecological Crisis" by Lynn White suggests that there is a link between religion and the environment. He says that the attitude of individuals towards nature must be changed. The environment must be respected and treated in a manner that will preserve it for future generations.

An individual's attitude towards religion largely depends upon what has been taught to him through his relgion. Sevier states that, "Traditionally religion used to play an integral role in linking people to the natural world, imbuing people with the knowledge and values that make caring for it a priority". When people respected their religions, they took care of the environ-

ment. But when people started moving away from religions, environmental problems started. For example, in the western countries where the religions had less influence, the environmental problems started worsening.
This concept is clearly explained by Carlos Parda (2008) as follows:

"The change of the worship of God to the worship of the Reason had to do with the emergence of the industrial revolution and the emergence of trends such as capitalism, communism, mercantilism, the development of technological power which has not been subject moral issues, and other factors that have been developed with the massive devastation of nature".

A spiritual world seems to be far better than the technological world. In the technically advanced world, people are becoming cultureless and does not seem to care aboour environment and what is it provides us. The modern society is going behind materialistic things that they forget to think about the consequences of their actionsn see people spending litres liters of water unwantedly although they know how they might suffer if there is water scarcity. They emit tons of Carbon dioxide even after knowinhe impact that it creates to on our planet. They produce waste continuously even though they are aware of the fact that they have to live with those wastes every day.

Figure 4.8: People Care Less About Environment.

Source: https://www.smallfootprintfamily.com/why-plastic-bottles-are-bad-for-the-environment

There is nothing wrong on the part of religion. Every religion has ethicalnd moral factors that teaches its followers to respect and protect the environment. No other force than religion can guide human beings to preserve the environment. The human population will listen to a higher and supernatural force than any awareness campaigns. So, religion is much needed to take our envient in the same manner to the future generations. In fact, religion is absolutely necessary.

Figure 4.9: Religion Teaches to Protect Environment.

Source: https://www.eoi.es/blogs/carloscerdan/2012/01/08/dp-4-relationship-between-religion-and-environmental-conservation/

Environment existed some 3500 million years before humans were created. Without nature, life is not possible on Earth. That is why every religion preaches the importance of nature.

Hinduism strongly believes in the interconnection between life and forces of nature. Mountains, trees, rivers, certain animals and birds are also sacred for them. Hve that every living thing that has life has a soul. According to Bhagavad Gita, the sacred book of Hindus, all life forms are inter-dependent on each other.

Figure 4.10: Hinduism Sees God In Everything.

Source: https://www.deviantart.com/yogeshvara/art/GANGA-MAYI-KI-JAI-34876094ristianity also emphasizes upon the fact that humans must not destroy the biological creation of God. In the sacred book of Christias, the Bible, it is stated that, "For that which befalleth the sons of men befalleth beasts . as the one dieth, so dieth the other. so that a man hath no pre-eminence above a beast." At many places in the Bible, there are messages on the preservation of trees and fruits, conservation of wildlife and agricultural lands.

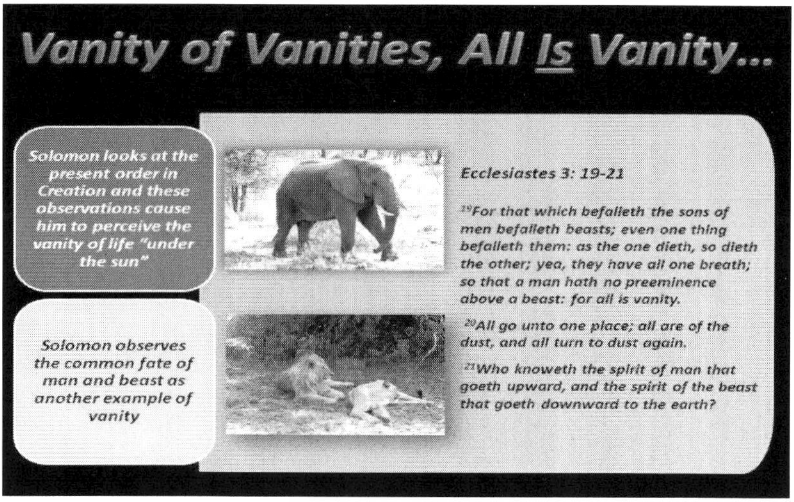

Figure 4.11: Belief of Christianity.

Source: https://slideplayer.com/slide/277450/

Islam believes that humans are the trustees of God who are entrusted with the work of safeguarding the universe and the living beings in it. The Koran, which is the acred book of Islam states that, "There is not an animal (that lives) on the Earth, nor a being that flies on its wings, but (forms part of) communities like you". Wasting natural resources is not liked by Muslims.

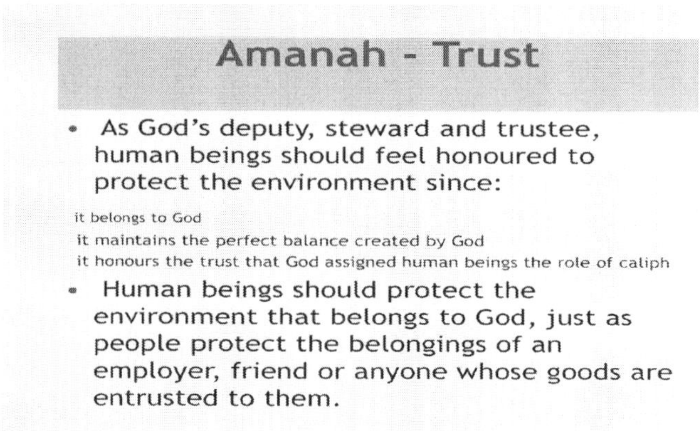

Figure 4.12: Islam and the Environment.

Source: https://slideplayer.com/slide/1221907/

Jainism believes in the inter-dependence of nature and human beings. This religion urges its followers to follow ahimsa (non-violence) not only towards other human beings but also to nature as well.

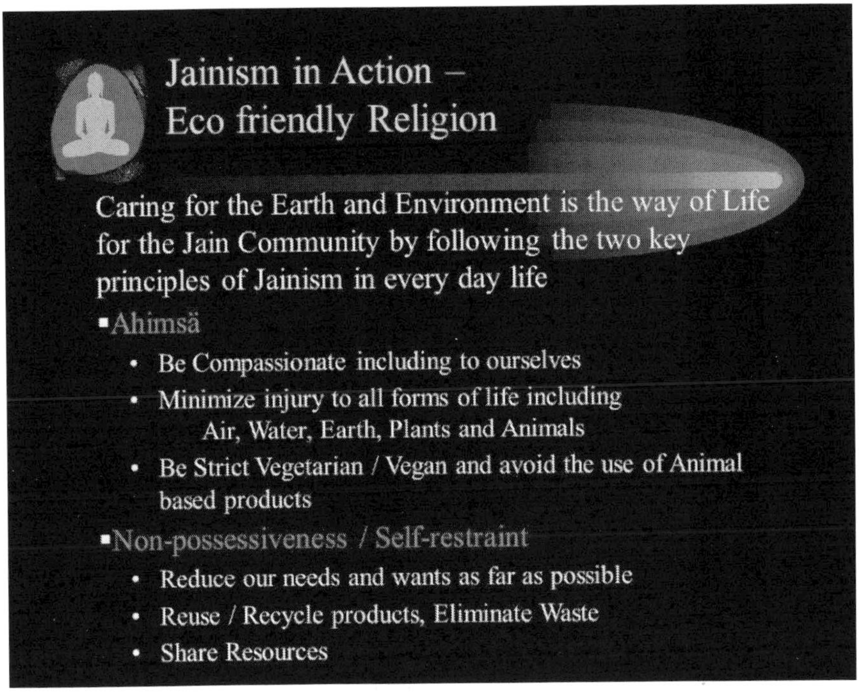

Figure 4.13: Jainism and Environment.

Source: https://slideplayer.com/slide/4207321/

Buddhism cares very much for nature and wildlife. It teaches t is respected when protection of biologthat we show for nature and that livingmone environment is veryas essentialvital.

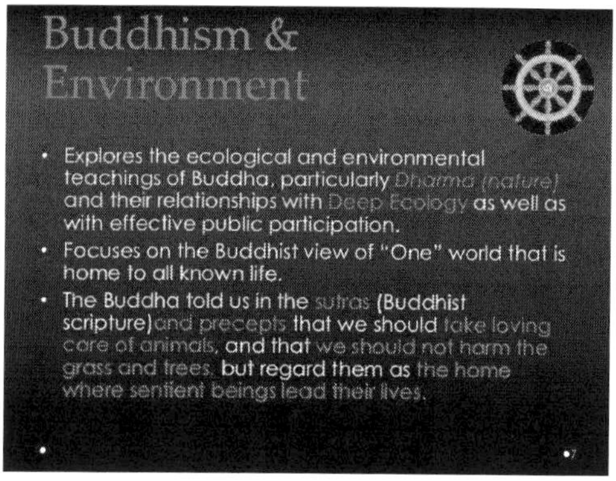

Figure 4.14: Buddhism and Environment.

Source: http://s3.amazonaws.com/brainfree/buddhist-teachings-on-the-environment.html

In Judaism, the Torah says, "When God created Adam, he showed him all the trees of the Garden of Eden and said to him: 'See my works, how lovely they are, how fine they are. All I have created, I created for you. Take care not to corrupt and destroy my universe, for if you destroy it, no one will come after you to put it right'". Thus Judaism clearly stresses on the conservation of nature and how we humans are connected with the environment.

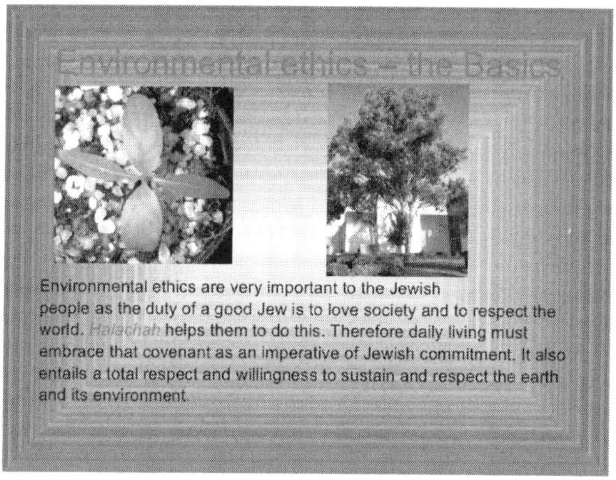

Figure 4.15: Judaism and Environment.

Source: https://www.slideshare.net/old_roofrat/jewish-environmental-ethics

The religious practices and ceremolowed in Shinto religion carries carry the message of the relationship between people and nature.

Figure 4.16: A Shinto Ritual.

Source: https://www.patternz.jp/shinto-beliefs-5-core-values/

All the above points clearly prove that all faiths share a common message i.e., living in harmony with nature. There aurch of Lebanon has protected theThe "forest hot spot" (a WWF Mediterranean Programme) or the forest of Harisa, has been protected bean Programme "forest hot spot", for over 1,000 years. In Thailand, in order to prevent logging and preserve the endangered forests, the Buddhist monks built small monasteries in the forests and made them sacred. Solarer has been harnesermany has installed solar power in 3ative w with the result that as many as 30 institutions doptinge inspiration of the local church.

The Sikh community in India has also launched an initiative to reduce the amount of fossil fuels used in the kitchens of their templ Kwanzaa, which has its origin in dating back to African harvest festivals is to enecological heritage of our planet.

Very often the leaders of the various igious group meet and discuss on how their religions are working towards the safety of the environment. Representatives of five major world religions – Christianity, Islam, Hinduism, Buddhism and Judaism were brought together by WWF during September 1986, to declare to the world how their religions care for nature.

This meeting ended up in the Assisi Declarations. After the meeting wasted, three more religions namelJainism and Sikhism, promulgated communicated declarations to accompany those of the other religions. Many such conferences and meetings are conducted periodically by religious groups and many new initiatives are taken for the welfare of the environment. Every religious scholar is eted to acquaint himself with the environmental problems.

During the 41st Cairo Climate Talks' panel discussion, representatives from different religion talked about the relationship between religion and the environment. H.E. Julius Georg Luy, German Ambassador to Egypt said that "Pollution and global warming pose an even greater threat than war," he said, "and the fight to preserve the environment could be the most positive way of bringing humanity together. During his speech, ambassador Luy expressed his thoughts about religion and said that since religion forms a very big part of Egyptian people, religion is very important to raise awareness on environmental issues. Dr. Saleh Ed-din Nefeily, a professor of English literature at Al-Azhar Universit was a member of the panel. He hads a deep knowledge of Islam and according to him, the Prophet's traditions are full of examples of his care for animals and plants. He explains that the environment that consists of water, food, flowers were created by God for us and God trusted human beings to take care of all his creatures. Dr. Saleh stre upon the fact that people who has have highly educated and if they still throw garbage out of the window, the only way to change their attitude is through religion and moral values taught by religion. They need to be educated about the harmful consequences of destroying what sustains life.

The third member of the panel was Father Angelos Guirguis who was the Pastor of the Cavern Church in Old Cairo. He has a different view on the relationship between nature and the humankind. He believes that "we need to reach a deeper understanding of our scriptures in order to stop viewing Nature as an object at our disposal, but to see humankind as an intrinsic part of Nature." Religion and environment support each other and are having a great connection from time immemorial and they will continue to have this relationship in future also.

4.3 RELIGION HELPS US TO BE ENVIRONMENTAL FRIENDLY

Reion plays a very important part in of everybody's life; it has a great influence on them. But the real question is that does this belief help them to be environmentally friendly. Does the belief in God make humans to take care of all the creatures and other things in the environment?

Let's see which religious folers are more friendly and caring to for the environment.

Compared to Christian fundamentalics are more willing to financially support the environment financially. This was identified from the fact that the Chrin countries where the Catholics was were in majority had more areas where nature was allowed to remain the same than other countries. There are many findings to prove the differences among the differen Christians. A 1993 study published by Andriest and sociologist Andrew Greeley looked at the extent to which environmental co1993 was spend on conserving the environment. The willingics was more than Andrew financially support the environment. This means that it is not important whether a person is Christian but what is more important is that what type of Christian is he because this influences his behavior towards nature.

Emhepheard-Walwyn while doing her PhD Ph.D. thesis at the University of Kent in Canterbury, UK, studied about how Kenyans felt about sacred sites. These places have both biological and spiritual significance and were created and maintained by cmunities of people arch Shephearlwyn found out thatShe writes, "someSome of the Christian people interviewed felt the forests should be destroyed as they are associated with the traditional faith, which they believe to be evil." Some Christians said that those places are the dwelling place of demons and superstition.

Figure 4.17: Sacred Mijikenda Kaya Forests, Kenya.

Source: https://www.pinterest.com/pin/1755700854451010/

Sometimes people converting to a religion from other faiths listen to the teachings of their religious leaders and act accordingly. So, if the religious leaders do not teach them wisely on how to be environmental friendly then definitely they'll be ignorample, a 2006 study published in 2006 by Leela Hazzah of Lion Guardians showth to become evangelical Christians hadill lions than tthat kept theirthe traditional faithpractionerays that these converted protestantsProtestants did not care much about national parks or wildlife. Since the Maasai's were not exposed to the outer world they relied much on their pastors for knowledge.

Thus, not only religion but even religious leaders play a very important role in protecting nature.

Buddhism has a good reputation on the conservatio n of the environment. Buddhism teaches people not to harm any animals. But this too is only apparent for Buddhists who are deeply religious. That is why conservationists need to integrate their ideas into religious thinking. Their message must differ according to the audience. Hinduism, Buddhism, and Islam already have great philosophies about nature and wildlife.

Some religious leaders do not accept science. They are of the opinion that science and religion do not mix up. But in reality, e religious ritual and practices has have scientific reasons behind it most of which depend upon the environment. If these religious leaders and conservationists are enlightened on the ways by which science can save or destroy the environment and how religious practices hints to it, then they can spread this message to a large group of their followers. There are many voluntary bodies and organizations that take ths as an initiative and work on it.

The Alliance for Religions and Cn (ARC) is a secular body that helpsto assist thn of environmental prers to create environmental programs based on thei beliefs and practices so that it canto reach the common man. Onecessful projects accomplished by ARC is on the island off the coast of Tanzania. Fishermen on the islands of Tanzania were using dynamite as a quick and easy method to catch maximum fish. But this method of fishing had serious drawback as they were damaging the corals and killed immature fish and turtles. The local environmental protection organizations tried to educate the fisherman on the negative impact of dynamite fishing, but there was no use. Even when the government passed laws to ban it, the fishermen did not take it seriously and continued their method.

It was then ARC came into the picture. The first thing that ARC noticed was that all the fishermen were Muslim and they wer greatly influenced by the local sheikhs. So, ARC understood that in order to make the fishermen understand their mistake it is necssary to first educate the local sheihs. The ARC officials showed the sheikhs passages from their sacred book Koran that promoted environmental safety and advised them that dynamite fishing is aainst the teachings of Koran. The sheikhs passed on the information to their community people and as these people were sincere to their religion they all listened and stopped dynamite fishing.

Figure 4.18: Dynamite Fishing.

Source: http://www.bbc.com/earth/story/20170206-religion-can-make-us-more-environmeendly-or-not

When one of the fisherman fishermen was interviewed by the Christian Science Monitor in 2007, he said, "“I’ve learned that the way I fished was destructive to the environment. This side of conservation isn’t from the mzungu [“white man” in Swahili], it’s from the Quran.”

The problem of dynamite fishing was not resolved by ARC alone. There was another organization which supported it, this organization was called the Islamic Foundation for Ecology and Environmental Sciences (IFEES) and Fazlun Khalid was the founder of the organization. After completing his degree in theology, Khalid concluded that Islam religion is basically about nature. So, in 1980, Fazlun Khalid started the IFEES organization as he was so passionate about nature and he also observed that Muslims had lost their connection with nature and even they started to go behind wealth. The sole purpose of establishing IFEES is to remind Muslims of the environmental values that the Quran teaches.

The IFEE initiatives such as . They have created an Islaable energy plants and recycling projects. "This eco-village was built based on the practices of the prophet on how to manage natural resources," says Khalid. Similarly, in Indonesia, IFEES is working with schools to restore the rainforests. Khalid believes that the religious movements are taking up new versions and they are much ready to embrace nature. "Faith-based organizations played a key role in the recent climate change negotiations, and IFEES were cornerstones in the creation of the Islamic Declaration on Climate Change," he says.

Figure 4.19: School Children Help in Restoring Rainforests With Support From IFEES.

Source: http://www.ifees.org.uk/indonesia-2/

This sort of approach has started working. A study conducted in Indonesia in 2013 showed that inculcating messages related to environmental protection into Islamic sermons increased awareness among people. So, the Indonesian government passed laws on Islam prohibiting illegal wildlife trafficking and poaching. Anything said with religion has greater power than the power of the authorized peotract a lot of people with its preaches and teachings and have the power to route hfor a common cause. Religions emphasizes on what we already have and what we are gifted with. Approaching people with religion will yield great results rather thanthem about scientific problems. Environment The environment is not what it was earlier; it is facing a lot of changes. The environment is dying. In order to solve this problem, propagating environmental awareness through religion is the only profitable way.

4.4 HOW DOES RELIGION BENEFIT AND PROTECT THE ENVIRONMENT?

Rrotects the environment through its believes beliefs and teachings. It sets up the moral framework of how an environment should be and how should it be treated. All the major religions such as Hinduism, Christianity, Islam, Buddhism and Jainism have a common approach to the environment i.e., conservation of the environment. All these religions have the below mentioned approaches towards religion:

- Human beings use natural resources according to their needs as they are on the top of God's creation
- Human beings are responsible and accountable for their use of natural resources
- The misbehavior of human beings leads to the disruption in nature
- Nature is sacred and nature is worshipped as God

Human beings use natural resources according to their needs as they are on the top of God's creation

This approach states thuman beings are the supreme force over all the creatures. Shey can control all other creations and use the natural resources for their needs. This phenomenon is misused by many and the environmental groups are really concerned over this attitude of humans. The actual way on how this approach works is that since humans are the top of God's creation, they have to safeguard all the other creations and use the natural resources only when it is required. Following such rule will benefit the environment.

Human beings are responsible and accountable for their use of natural resources

This approach does not show human beings as superior in the creation list but as a part of nature and that they are responsible to protect the environment. This approach teaches every human being that they are answerable for all their actions towards the environment.

The misbehavior of human beings leads to the disruption in nature

This approach is kind of a warning towards human beings. According to this approach, our unmindful actions and behavior may cause huge damage to the environment and this affects even the human community as a whole. So this approach askman beings to think before they do something to the nature.

Nature is sacred and nature is worshipped as God

It is not only enough that we worship nature but it is very important that we take care of nature. Each religion through its beliefs expresses concern for the environment and guides its believers towards conservation.

Christianity promotes environmental care through the teachings of the Bible. It respects nature's divinity and accepts the responsibility of humans to look after God's creation. Even Islam follows the same principles when it comes to the environment. They preach them through the Quran that refers to nature as bered.

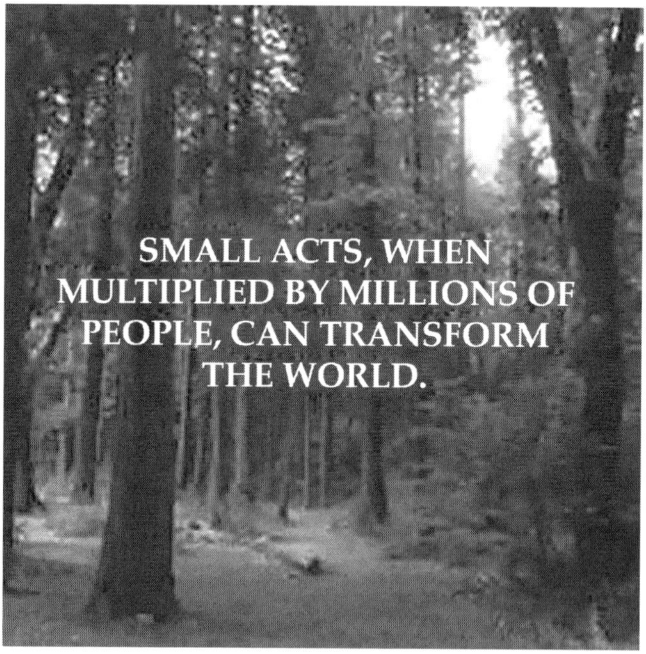

Figure 4.20: Buddhism Preaches Humanity.

Source: https://ecolifeke.com/role-religion-environmental-conservation/

The Buddhists believe in oneness and connection between humce taking care of nature is taking yoursectice is compassion and that of peace and comng a balance between human practices and conservation.

Hinduism is also in the club of protecting the ecosystem. Hindus follow this in their life with sincerity. Mahatma Gandhi, w is celebrated in India as the father of Nation is a the best example to prove how he incorporated his religious values into his life. Mahatma Gandhi lived in peace and harmony with his fellow human beings and nature too. He led a very simple life or said as an ecological treatise. He lived as an example and role models entire life can be said as an ecological treatise. His daily routine of having a small meal of nut and fruits in the morning, the exercises that he practices every day, his observances of silence periodically, his morning walks, his spinning wheel, his values for truth, accepting non-violence, celibacy, fasting and his abhorrence of waste – everything was so true to his religion. He inspired many of his contemporary religious leaders and environmental leaders in their effort to protect the environment.

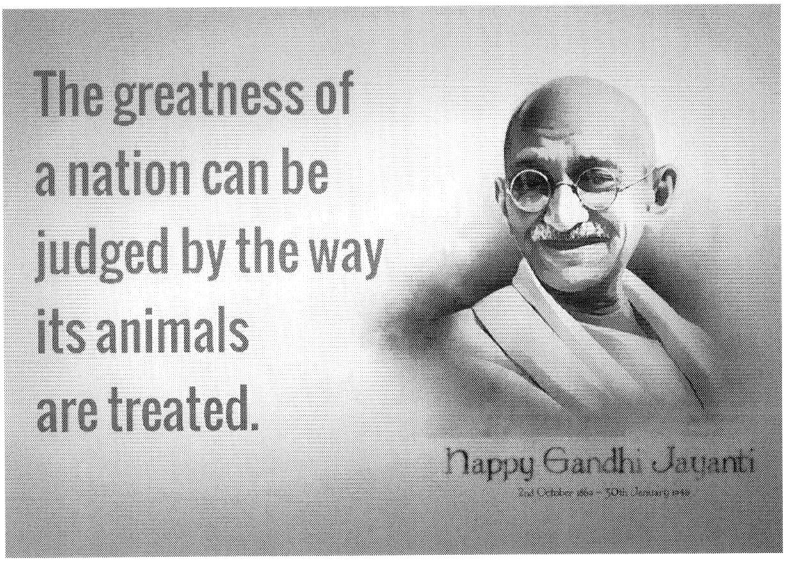

Figure 4.21: Mahatma Gandhi.

Source: http://www.picturequotes.com/the-good-man-is-the-friend-of-all-living-things-quote-1811

Judaism believes in the fact that nature is central and God is the owner of it and no one has the right to alhould leave nature as how it is. Judaism disc not support unnecessary destruction and destructive via human activities. Judaism helps its followers to aintain a proper environment through its teachings.

AAll rious groups are actively presenting themselves in the environmental protection activities. The role of religions in protecting the environment has led to the development of a new field alled the eco-theology. Eco-theology includes orporates the concept of ecosystem management in religious teachings. Even hough new concepts like these arises and religions takes so many efforts to live in accordance with nature and protect nature, all these will give he desired resultcan still l force for conservation today: a religious association association functions to restore sacred cente plant diversity close to that of a ny nature reserve. Many traditional societprotect the biodiversity at sacred sites by limiting the human activity. Liu Hongmao, Director of the Chinese Academy of Sciences' Xishuangbanna Tropical Botanical Garden in Menglun, People's Republic of China, and three coauthors in theApril issue of Biodiversity and Conservation states that, "Conserving biodiversity based on cultural and religious values is often more sustainable than based only on legislation or regulation".

Figure 4.22: Xishuangbanna Tropical Botanical Garden.

Source: https://www.discoveryloyalty.com/Anantara/Anantara-Xishuang-banna-Resort/Xishuangbanna-Tropical-Botanical-Garden-Night-Tour

Liu along with his colleagues explored the role played by religion in conserving plants at Xishuangbanna which is a region in China that accounts for only 0.2 % of the country's land but has about 18 % of its 30,000 plant species. This regions tropical forests were decreased in half since 1950 and about 600 secies of plants have gone extinct there. Dai people anic group people of Xishuanna i people wiefs traditionally played an important role in conservingto conserve local plant diversity.

According to Dai people's polytheistic beliefs, every villae had a holy hill forest where they believed that the God's lived and so hunting, gathering, and cutthibited in these villaregion has about 250 holy hills , each of which is about 1,000 -1,500 hectares each. 15 protect68 plant species When thnder abou the holy hills and say that these holy hills cato hence n be considered as small nature reserves established by the Dai traditional beliefs.

That was the polytheistic part of it. According to the Buddhist beliefs, each village must have a temple with a garden containing plants related to the religion, such as the flowers and fruits used as offerings during ceremonies. Out of the 558 temple gardens in Xishuangbanna, Liu and his colleagues surveyed 51 plants and found out more than 100 plant species. Liu and his colleagues confer that the temple gardens have played a significant role in the conservation of plant species.

Figure 4.23: Menghan - Dai Temple.

Source: https://global-geography.org/af/Geography/sia/China/Pictures/Xi-shuangbannnghan_-_Dai_Temple_1

TThe researchers understood that the traditional beliefs are helping in preg the plant species that led to the establishment of . So, the AssociationTraditie role of traditional beliefs in conserving plant diversity. Many courses and services to conserve these sacred sites were undertaken as weat six temple and four holy hill sites which were destroyed.

These hill sites were restored and when Liu and his colleagues compared the holy hill sites before and after restoration and found that plant diversity increased by more than 8 percent, approaching that of the nature reserve. After continuous researchk at these sites, Liu and his colleagues believed that the combining traditional beliefs with conservation can be widely used and it will be the most successful approach.

"All of the world's major religions are sensitive to the importance of biodiversity and nature,"

Robin Meadows

A community in northern Ghana had preserved a small forest for nearly three centuries as it was believed to house a local spirit. Although they were threatened over the years fearing activities such as road-building, mining, and electric-pole installation, this sacred forest was somehow preserved by the localdecentralized government and local protection of ecosystemsThis shows the importance of religion in the maintenance of diversity.

There are many similar references to the protection and worshipping of the wild in Hinduism. Vedas, Upanishads, Puranas and Sutras are all the sacred text followed by Hindus and all these contain information about it. The Chipko (tree-hugging) Movement is the most widely known exaHindu environmental leadership, but there are also century century-old examples of Hindu action for the weevere their rivers, mountains, trees, animals and the earth.

Figure 4.24: The Chipko Conservation Movement Practiced by Hindus.

Source: https://twitter.com/heonggallery/status/756132032154001412

Hinduism is remarkably the only religion that has diverse religious and cultural phenomenon. Hinduism has different theologies and few are listed below:

A

- The way we treat nature directly affects our karma.
- The Earth is a manifestation of the Goddess and so should be treated with respect.
- Air, Water, Fire, Space and Earth are the fiveelements of nature and these are the foundation of life.
- TTaking caring the environment is an important expression of harma.
- Thus,vingis a model for the development of sustainable economies.

In the past centuries, Indian communities similar to the traditional Dai people in China did not have a separate understanding of nature but they were very religious. Rurities such as Bhils, Bishnois and Swadhyaya have maintainedinvolve strong communal practices to protect local ecosystems such as forests and water sources. They are ignorant of the fact that their activities are protecting the environment; they do all these only because they const as their dharma as seen with protection of trees by . The Bishnois protect animals and trees; Vrikshamandiras (tree temples) tes) of the Swadhyayis build Vrikshamandiras (tree temples) and Nirmal Nirs (water harvesting sites). They knew that God's creation must be treated with respect.

Figure 4.25: A portrait of Amrita Devi and her daughters, members of Bishnoi Community who laid down their lives in 1730 to save trees from being cut by the Maharaja of Jodhpur.

Source: https://www.tribuneindia.com/news/punjab/tributes-paid-to-bishnois-who-died-protecting-trees/459748.html

Most of the Hindus fear the fact that their actions towards the environment will affect their Karma. According to Hinduism, Karma means that for every good and bad action of a person there is an equal consequence which constitutes the Karma and the fate of our future. Hindus believe that it is our Karma that decides what our next birth will be and how it will be. Good moral behavior and good activities constitute good Karma and bad actions create bad Karma. Even though a person has done harm to the environment, he after realizing the mistake can opt to protect the environment in the future.

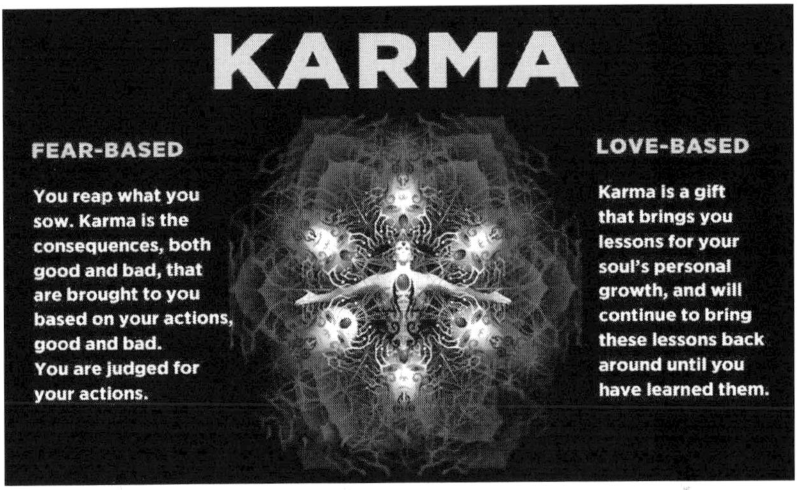

Figure 4.26: What is Karma?

Source: https://www.sanskritimagazine.com/spirituality/karma-love-based-fear-based/

Hinduism is enriched with believes that respect nature and worship Earth as Devi, the goddess. Since Earth gives everything that the human needs, they return their gratitude by protecting it. Even the day-to-day activities that the Hindus do at home benefits the environment and it is a real tribute. Before getting out of bed every morning, most Hindus touch the floor and pray to the goddess to forgive them since they are going to trample on her body i.e., the Earth's surface. They decorate their doorways with Kolams (small or big artwork) of rice in the morning. Drawing Kolams is an expression of Hindus to offer sustenance to the Earth, just as the Earth sustains themselves. Also since Kolam is drawn using rice powder, small insects, ants, crows, and squirrels can feed on it. It is a way of feeding other creatures of God.

Figure 4.27: Kolam Ritual Followed by Hindus Every Morning.

Source: http://www.dsource.in/resource/rangoli/types-rangolis/olam

Yoga and meditation that are taught in Hinduism proves that the Earth consists of divine energy and through yoga and meditation people can connect to this divine enerThis divine energy gives peace of mind, cures diseases, givegives us the courage and unites people. Teacheave used these teachingsto demoxploitation of the environment, women and indigenous peoples.

Hindus believe in Karmas which means that they also believe in reincarnations. Reincarnation or rebirth gives a sense of interconnecl creations. Based on the previous karma, They believe that every being travels througiffnimal, bird or another part of the wider community of life. Since everyone has to pass so many lives of reincarnation in order to attain liberation, reincarnation creates a sense of solidarity between people and all living things.

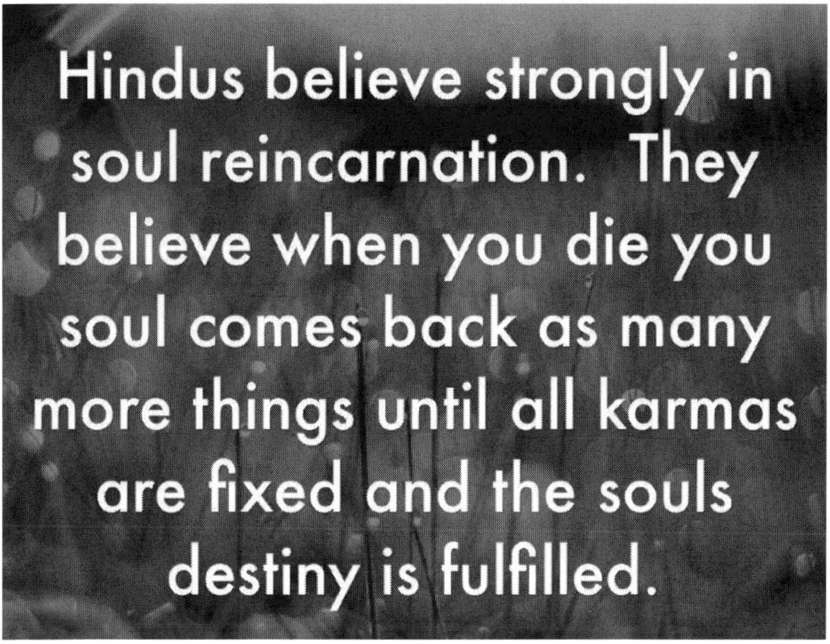

Figure 4.28: Definition of Reincarnation.

Source: https://www.haikudeck.com/hinduism--education-presentation-40UX8FH0nw

Hindus also believe ahimsa (non-violence) to be the greatest of all dharmas and ahimsa towards nature and other creatures will improve one's Karma. Hunting of animals or even harming them is considered ahimsa and doing so will affect one's Karma badly and hold them from attaining moksha i.e., the ultimate liberation. In order to prevent themselves from getting into bad Karma, Hindus are advised to avoid such activities and follow a vegetarian diet instead of killing of animals, birds and fish.

There is a well-known Hindu teaching which goes as, "Tain tyakten bhunjitha" which means "Take what you need for your sustenance without a sense of entitlement or ownership." This is the reason why many adopt to Sanyasa (asceticism). Sanyasis live a simple life and restrain themselves from over- consumption of anything and Sanyasa is sought to be a path to liberation.

4.5 HOW DOES ENVIRONMENT SUPPORT THE DEVELOPMENT OF SOCIETY?

Let anything and everything happen in the society – let there be economic crisis, wars, and any other social problems but the significance given to the environment will still remain the same. The entire life support system of humanity depends on the well-being of all the species living on Earth. A Russian scientist named Vladimir Vernadsky in 1920sphere' that is an interalpendent. Within the overall biosphere or ecosystem, there areThe presence of sm rainforests, etc . are seenoceans, the desert and the tundra. This entire biosphere or the ecosystems benefit humanity and also support the development of the society.

A scientific study conducted in 2016 proves that there are more interconnections between species, their size, and the ecosystem. Existence of biodiversity has many benefits such as –

- Purification of water
- Rain Harvesting
- Natral cycling of elements
- Fertility of Soil
- Resisting Drought

Our ecosystem provides several natural resources that are useful for people hence, called as "provisioning sernd the natural resourcesin the environment benefit the society:

Wood

According to TEEB, wood that is derived from forests or plantations is used as fuel or in carving furnitures.

Water

Water is declared as a human right by the United Nations and this shows how much important water is. Nature has given us such an esential thing in abundance for us.

Medicines

TEEB states that, many plants have been used as medicine for more than hundreds of years. Even today the modern pharmaceutical companies are exploiting them and using them to prepare medicines.

Clothes

Our environment has many plants, trees and animals that help us wear some of the best clothes. Wood pulp, jute, cotton, hemp and anioducts like silk, wool, leather are considered as natural fibresfibers afibers. Synthetic clothes are produced from petroleum products.

Biofuels

Nature h in producing biofuels. For example, biofuel like bioethanol are is extracted from wheat, corn or biomass crops like willow, says, Union of Concerned Scientists.

Clear Air Pollution

Our environment consists of hundreds and thousands of trees that help in cleaning the air. Also, they regulate the carbon cycle by reducing its concentration in air by inhaling it. Trees produce oxyen which is needed for humans to breathe.

Control Temperature

AAccording to TEEB, plant and trees provide shade an help in controlling the heat during summer.

Prevent Disasters

Our ecosystems when left as it is, has the ability to limit the occurral disasters as g and its associated destruction of habitats and humansettlements.

Entertainment

People enjoy the real beauty of nature. Many natural places attract people and these places become places of recreation from which tourism flourishes. Even the natural occurrences help tourism sports such as skiing, rafting etc., Nature is considered necessary for proper physical and mental health too, according to the University of Minnesota.

Food Chain

Food chain is the best example to prove the benefits of our ecosystem. Encyclopedia Britannica explains the food chain process as – plants use sunlight, water, air and other elements in the soil to prepare food for themselves and these plants are in turn eaten by animalt of the environment without whom the food chain would be broken. In the pyramid of the food chain, humans are on the top of it since they eat plants as well as animals as their food.

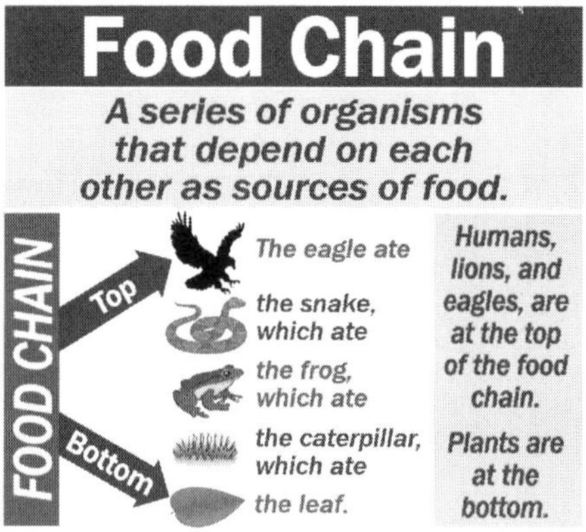

Figure 4.29: Food Chain.

Source: https://marketbusinessnews.com/financial-glossary/food-chain/

An international scientific study conducted in 2012 shows that the go and services that are obtained from our environment worldwide is are estimated to be of worth $125 trillion per year. The real value of the environment is much more than thmonetary benefit of it. The environment provides everything that a society needs and helps in its development. It is just that we have to maintain the functionality of the environment to derive its benefits.

REFERENCES

1. http://www.businessdictionary.com/definition/environment.html
2. http://www.yourarticlelibrary.com/environment/meaning-definition-and-components-of-environment/6157
3. https://simple.wikipedia.org/wiki/Environment
4. https://www.tutorvista.com/content/science/science-ii/environment/introduction.php
5. https://www.factmonster.com/science/environment/what-environment
6. https://www.soas.ac.uk/cedep-demos/000_P500_ESM_K3736-Demo/unit1/page_08.htm
7. http://www.bbc.com/earth/story/20170206-religion-can-make-us-more-environmentally-friendly-or-not
8. https://www.eoi.es/blogs/carloscerdan/2012/01/08/dp-4-relationship-between-religion-and-environmental-conservation/
9. https://iefworld.org/unepgc21.htm
10. http://cairoclimatetalks.net/events/religion-and-climate-change-relationship-between-god-and-environment
11. http://hollys7.tripod.com/religionandecology/
12. https://www.slideshare.net/elove/religion-and-the-environment
13. https://ecolifeke.com/role-religion-environmental-conservation/
14. https://www.conservationmagazine.org/2008/07/religion-can-benefit-conservation/
15. https://www.wri.org/publication/religious-beliefs-and-environmental-protection
16. https://www.huffingtonpost.com/pankaj-jain-phd/10-hindu-environmental-te_b_846245.html
17. https://greenliving.lovetoknow.com/why-is-conserving-energy-important
18. https://www.routledge.com/Religion-and-the-Environment/Gottlieb/p/book/9780415554510
19. http://www.sustainablebabysteps.com/religion-environment.html
20. http://www.sustainablebabysteps.com/religion-environment.html
21. https://www.theglobeandmail.com/arts/books-and-media/science-religion-and-the-environment/article1205431/

22. https://www.palgrave.com/gp/book/9780333919743
23. https://en.wikipedia.org/wiki/Religion_and_environmentalism
24. https://www.bloomsbury.com/uk/religion-in-environmental-and-climate-change-9781441169297/

Religious Practices That Affect the Environment

5

"If people are to know and understand the real world, they must give up superstitious beliefs because they have a narcotic effect on the mind."

–Karl Marx

CONTENTS

Being devoted to religion is good. Following religious principles, walking in the footsteps of religious gurus are well appreciated. But everything is fair only until it makes sense. Nowadays, people do many unwanted things in the name of religion. They follow many superstitions, practice many rituals, believe ancient myths and adapt them in their life. These don't seem to be good and this will lead to downfall. This Chapter is all about different practices that human beings follow in the name of religion. Most of these practices affect our environment also.

5.1 IMPORTANCE OF RITUALS

A ritual is a sequence of activities involving gestures, words, and objects, performed in a sequestered place, and performed according to set sequence. Rituals can be in accordance to traditions of a community, such as a religious one. Rituals may entail the involvement of rules or society or traditions or formalism, and performance.

Rituals are seen in known human societies. They include worship rites and sacraments of organized religions and cults, along with rites of passage, atonement and purification rites, oaths of allegiance, dedication ceremonies, coming of age ceremony or rites, coronations and presidential inaugurations, marriages and funerals, school "rush" traditions and graduations, club meetings, sporting events, Halloween parties, veterans parades, Christmas shopping and more. Rituals are partly seen in scientific symposia, jury trials, execution of criminals, and even hand-shaking and saying "hello". In terms of spirituality, rituals are vital for providing a center for comfort. Examples are: Rosary, Gregorian chants, native and the Sufi dervishes dances that connect one with the source. The importance of rituals is on the same page as experiments in science.

'just as it is impious to ill-treat the statues of the Gods, in the same way it is not righteous to err regarding names.''

- Proclus

"He[simplicius] spells out what should be determinative for our attitude toward the Gods explicitly in three theses:

the Gods exist, they exercise forethought, and they do so in a way that is just and in accordance with correct reason."

The rituals must be done carefully and not in haste.

Rituals help in moving, motivating and connecting humans through expressed emotions to maintain identity. Examples are gay parades or love celebrations in Sweden or piercing in Thailand or High Mass.

They play a vital role in our mental, emotional and spiritual wellbeing. Dally Messenger, author of the book Ceremonies and Celebrations says, "While there has been a move away from religious ceremonies, people still need a ritual to mark major points in their life,". "Ceremonies are an expression of culture," he adds, "mechanisms which express and generate love, forge and declare the bond between individuals, and establish and identify the community."

Rituals also involve getting together as seen with Bastille Day celebrations in France or ANZAC day of Australia to bind people in identity Other rituals include reading a paper, a beverage or kisses or playing games or watching specific programs that all aid fullness and organizations in life. Lighting candles or praying before meals or celebrating functions together or welcoming babies that all are rituals. Apart from religion, ritual practices between couples such as intimacy and trust are important. Even self-help activities such as relaxations or waling or bathing can help bring peace amidst stress.

And, of course, AN

Certain rituals are performed without realization that allows for connection with the divine. The identity and basis of religion are seen in its rituals. The practice of observation is allowed for in rituals as seen with eating at correct times or dipping in a river or communion with spirits. While in religion, rituals allow for connecting with the divine, activities done daily too are rituals that maintain order and purpose. The observance of rituals can help in gaining knowledge of another's way of life. Rituals can be in public or in private involving singing or weeping or bathing in rivers or fasting or celebrations.

5.2 DIFFERENCE BETWEEN RITUALS AND SUPERSTITIONS

Each culture has its own myth and its own pantheon as to how the cosmos got originated and manifested in its various forms. Even though the theory is the same, the Gods in culture show variation. As seen in the God as male and female as Pistis Sophia and the Holy Father or Savithru and Gayatri or Samanthabhadra and Samantabhadri, respectively,

The fact that different cultures have discovered these myths on their own it is not wise to dogmatically assert that only one pantheon represents the manifested reality out there and the other pantheon is false or even to say the different cultures are talking about the same pantheon, instead it should be realized that not only the rituals of our nation work and the Gods of our pantheon are true but also it must be realized that the rituals of different cultures and their pantheon works indeed which philosophically we can term it as the divine Logos the working principle of the cosmos while simultaneously preserving and giving importance to the specific rituals and specific Gods discovered by each culture because only through this path of the Gods we can transcend the manifested reality and achieve the non-dualistic unity.

Rituals are varying that can evoke different reactions; the Satere-Mawe tribe of the Amazon observe the use of bulletants to look at coming of age that can be shocking to another community. Tattooing of the mouth is seen in the West African Fulani Tribehave women. Such rituals mark a shift to adulthood and strengthen ties among the community members.

Rituals allow to "perform at our best when we need to, stay calm when we're under intense pressure, and create a sense of balance in our lives", says Chris Attwood, co-author of the book Your Hidden Riches. Player Serena Williams dons the same socks or bounces the ball five times on her first serve and twice on her second which again are rituals observed to have a mindset for success. Studies show that winning is enhanced when one uses a "lucky" ball! Actors like Colin Farrel too follow rituals of particular boxers to the day of the shoot!

Superstitions are seen in human society that can damage the mindset as seen in the statement of the 1972 hit Superstition, associated with financial risk-taking and gambling behaviors."" No matter how advanced and rational we might like to think we are, superstition is a practice that remains widespread and has an impact – often detrimental – on our economic lives". For example, the 13th is avoided for decisions or buildings lack a 13th floor. The presence of horoscopes in papers too is a testament to superstitions. The numbers 8 and 4 hold importance for the Chinese community of Singapore. In Italy and Brazil, the number 17 is unlucky making airlines skip this number.

As 8 is lucky in China, the timing of the Beijing Olympics was on the 8th month of 2008 at precisely 8:08 pm! Such beliefs according to research impact new sales of houses or the numbers of floors of the houses bought. However, whether these beliefs actually have an impact on life is subject to controversy!

In the case of the Malagasy, non-venomous fandrefiala snakes are dreaded because of their red tails associated with shapeshifting. Hyenas are revered in Kombolcha, Ethiopia as they protect from evil spirits. The latter can be looked as an intermingling of rituals with conservation.

However, some rituals can oppose conservation efforts. This is seen with killing endangered aye-aye lemur as it was associated with death in Madagascar or owls in some Caribbean islands. Another example of conservation is the critically endangered radiated tortoise that is protected in Malagasy taboo that however does not apply to poachers or outsiders.

Figure 5.1: Radiated Tortoises in Southern Madagascar.

Source: https://www.atlasobscura.com/articles/endangered-animals-cultural-myths

Other examples of superstitions are warding off the evil eye. Such superstitions that are not placed on science but habit and are seen in cultures. Such belief is transmitted across cultures and the one involving lower costs are preferred.

Superstition Across Cultures

Certain examples include those who read horoscopes that however as mentioned above can impact worker motivation and business .

1. Russian house spirits any Russians are uncomfortable with the idea of shaking hands across a doorway where the house spirit resides.
2. Friday 13th:

   Not much is launched on this day. The origin of this is probably the day of the Crucifixion or the Knights Templar death or the Last Supper. However, this day does not deter travelers or those looking at hotels .
3. Hindus respect their ancestors

   This is the 16-day Pitru Paksha in India when products aren't launched.
4. Eclipse

   The Harvard Business Review found slumps during eclipses.
5. Icelandic elves

   Giant boulders are considered sites where elves live that are not touched upon by local construction workers.
6. Geomancers

   This refers to feng shui seen in even the Bank of China and the HSBC headquarters and product launches of top brands.

A simple example is to bless when one sneezes though one may not be superstitious.

What is superstition?

While there is not yet a coherent definition of superstition, belief in supernatural forces is an explanation. There is an involvement of individual beliefs and experiences.

Psychologists state that superstitious behaviors help people handle anxiety or stress such as the use of good luck charms during stress. An example is increased superstition during the Economic crisis in Germany in 1918 and 1940.

Figure 5.2: Black cats are less likely to be adopted.

Source: https://www.popsci.com/science-superstition#page-2

Other examples include certain clothes, charms, specific colors and numbers. Numbers come into play when number plates are looked at. Even certain places or things are considered cursed such as the Curse of the Pharaohs or Annabelle the Doll.

Figure 5.3: Dayton baseball players try to bring good luck by twirling their fingers.

Source: https://www.popsci.com/science-superstition#page-4

Rafael Nadal follows certain rituals of showers or water bottles; North Carolina shorts worn by basketball legend Michael Jordan that all are superstitious behaviors. Rafael Nadal has an array of rituals that he performs each time he plays. These include the manner in which he places his water bottles and taking freezing cold showers. Nadal believes these rituals help him to find focus, flow and perform well.

What all this shows is that superstitions can provide reassurance and can help to reduce anxiety in some people. But while this may well be true, research has shown that actions associated with superstitions can also become self-reinforcing in that the behavior develops into a habit and failure to perform the ritual can actually result in anxiety. This is even though the actual outcome of an event or situation is still dependent on known factors rather than unknown supernatural forces; a notion consistent with the often quoted maxim, "the harder you work (practice) the luckier you get."

5.3 SACRIFICES WITH NO MEANING

Sacrifice is the offering of food, objects or the lives of animals or humans to a higher purpose, in particular, divine beings, as an act of worship or propitiation. The term offering (Latin oblatio) can also refer to beverages in place of blood.

There are several aspects to this such as the the Germanic blōtan or Greek thusia or the Indic yajna or the Slavic żertwa while the Latin word sacrificium involves performing sacred activities . Metaphorically, sacrifice can also mean doing good or giving up something to achieve a greater good.

Animal sacrifice

Animal sacrifice is the ritual killing and offering of an animal usually as part of a religious ritual or to appease or maintain favor with a deity. This was ubiquitously seen in the Ancient Near East and throughout Europe and the Ancient Near East with its practices still on in several parts.

In certain cultures, only certain parts were given to Gods as seen in the ancient and modern Greeks while others burnt an entire animal (holocaust) as seen in the ancient Hebrews. This is different from slaughtering animals for food.

Figure 5.4: Sacrifice of a Pig in Ancient Greece.

Source: https://en.wikipedia.org/wiki/Animal_sacrifice

According to Walter Burkert, a scholar of sacrifice, hunting was the source of Greek sacrifices. This was a repudiation of killing the animals. An example is including the Dipolieia in Athens: the whole community to be involved in ox sacrifices against one person offering an entire sacrifice.

Human sacrifice

This involved pleasing a deity seen on many occasions and in many cultures (The Aztec or Pre-Columbian civilizations or sacrificing captured Conquistadores in the Spanish war). Examples included:

- Sacrifice when a king or high priest died to accompany the dead
- Natural calamities were considered angry gods who were appeased by human sacrifices.

This practice is not followed now anymore.

This involves the sacrifice of a sheep, a goat, a camel, or a cow that is healthy and conscious. "Therefore to the Lord turn in Prayer and Sacrifice." (Surat Al-Kawthar) Quran, 108.2 Qurban is an Islamic prescription for the affluent to share their good fortune with the needy in the community. (Qur'an 22:37: "It is not their meat nor their blood that reaches God. It is your piety that reaches Him...”). The meat is divided and is shared with the family and the poor.

Other examples are the pollution of the environment seen with rituals, an example is bathing, drinking, or having their ashes scattered despite considering the Ganges River holy by the Hindus. This has led to high levels of pollution.

5.4 HISTORICAL BELIEVES AND MYTHS STILL CONTINUED

Our ancient ancestors used the bond of ritual to create ties of kinship necessary for survival in a world rife with dangers. Early worship can be traced back from structure and hierarchy seen with rituals. An early record is the Tsodilo Hills, Botswana where sacrificial offerings of spearheads were made to a stone python, 70,000 years ago believed to be the ancestor of man.

With the emergence of modern religions, the consolidation of practices was seen through sharing and rituals. The American Thanksgiving dinner allow for connection to ancient history and the historical trend. Patrick Nunn, a geoscientist from the University of the Sunshine Coast in Australia conveys that the old mythological stories represent what ancient people observed. The creation of legends can instill a feeling of fear and responsibility towards the planet.

There is no historical proof as to which came first – is it the natural calamity or the story. No one can clearly say whether it actually happened. But these tales help us in sticking to our roots and even helps scientists in understanding the mysterious gaps that science cannot fill.

The Ark of Noah

Many of us would have heard about the ark of Noah. This is a famous story told in the households of Christians, Muslims and Jews and of course, the movie taken based on this story has been widely viewed by people from all across the world. According to this legendary story, God decided to destroy the Earth and he opted to do it by causing a flood. But God also decided to let live one man named Noah and his family.

God commanded Noah to build a huge boat i.e., an ark and asked Noah to fill the ark with a pair of every animal and species that lived on this planet. And finally, when the flood came, everything in this world got destroyed except the ark which had Noah, his family and the animals present in the ark. Whatever was left was the reason for re-growing the Earth. This is the story of Noah.

Seeing this story from a scientific point of view, the scientists say that there's not much water present to drown the whole of Earth as told in the story. There are also some other viewpoints that Noah's story may be based on other similar occurrences that had happened in the past. There were times that some piece of land was completely drowned inside water. There are also some geologists who say that Noah's story may be derived from catastrophic flooding that happened in the Black Sea during 5000 B.C.

Adrienne Mayor, a historian of Ancient Science from Stanford University says that in general people have the tendency to overstate what they know and cause more fear in one who listens. He says that this global flood is one such exaggeration but in reality, the movement of tectonic plates in the ocean are responsible for high elevation of water level which causes floods and tsunamis.

The story may be imaginary; we are not sure but the real intention behind the story is that humans should know that when they cause harm to one another and forget their responsibilities, God will definitely punish them.

Figure 5.5: The Ark of Noah.

Source: https://www.smithsonianmag.com/science-nature/ten-ancient-stories-and-geological-events-may-have-inspired-them-180950347/

The Oracle at Delphi

This is also one of the mythological stories told in ancient Greece. As the story goes, in ancient Greece, there was a town named Delphi on the slopes of Mount Parnassus and there was a temple to God Apollo in that mountain. The temple had a secret chamber in which a priestess named Pythia did some magical thing. The crack in the mountain rocks used to emanate sett-smelling vapors and Pythia would smell these vapors and go into a frenzied state and start speaking gibberish channeling God Apollo. The gibberish uttered by Pythia would then be translated into prophecies by a priest.

Scientists could not find any clue to confirm that the story is real. But the excavations proved that the temple in the story was real and there was gas emanating from the cracks and the gas may be benzene, ethylene or a mix of carbon dioxide and methane.

Atlantis

Atlantis here represents civilization and not a person. More about this civilization was told to the world by an ancient Greek philosopher called Plato. According to Plato, Atlantis was founded by people who were half human and half God and they lived in Utopia. People of Atlantis were said to have strong naval power. It is said that their houses were shaped like a series of concentric circles on islands and these islands were destroyed in a natural calamity. There is no evidence to substantiate that Atlantis was a real place. Scientists believe that many other civilizations that existed years ago might have inspired this tale. One such was the Santorini in Greece existed before a thousand years which was a volcano and island named Thera. It is said that this volcano erupted some 3500 years ago and this was a major disaster in history causing great damage to the island and the after effects of it continued for very long years. This disaster led to the decline of Minoans who were the dominating race in the Mediterranean region. The scientists feel that the story of Santorini would have inspired the creation of Atlantis. There are also many other inspirations like the City of Helike in Greece and the ancient metropolis which got vanished due to earthquake and tsunami during 373 B.C.

Pele, Goddess of Kilauea

This is one of the old stories and it is very dramatic compared to many other mythological stories. It is about Pele who came along with her sisters and relatives to Hawaii. Initially, she went to Kilauea and it was there she met

Lohi'au. Although Pele liked him, she did not stay there because the weather was not that warm. So, Pele went to the crater at Kilauea; she asked her sister Hi'iaka to bring Lohi'au to her and in return, Hi'iaka asked her sister Pele to not destroy her forest in her absence.

Pele gave forty days' time to her sister to accomplish the task but since Hi'iaka did not return within the stipulated time, Pele thought that Hi'iaka had betrayed her and took her love and so she set the forest on fire. When Hi'iaka heard that her sister Pele had destroyed the forest, she made love to Lohi'au as revenge against Pele. Agitated in anger, Pele killed Lohi'au and threw him into the crater. In order to recover the body, Hi'aka dug furiously that even rocks flew and finally she recovered Lohi'au's body and they are together now.

This is the great mystical story of Pele.

Figure 5.6: Pele Considered as Goddess of Kilauea, Hawaii.

Source: https://www.smithsonianmag.com/science-nature/ten-ancient-stories-and-geological-events-may-have-inspired-them-180950347/

According to scientists, the story is a representation of the high volcanic activity at Kilauea. During the 15th century, there was a continuous flow of lava for sixty years covering nearly 430 square kilometers of the Hawaiian island and this is compared with the forest fire talked about in the story. This was even seconded by Donald A. Swanson, a volcano scientist in the U.S. Geological Survey as he says that this lava flow was the only possibility to impact such a large area. Even though we are uncertain about the legitimacy of the story, everything said in the story had some connections to the place. The formation of the volcano's modern caldera represents the furious digging on the crater by Hi'iaka.

The Bridge Built by the Hindu God Rama

The Ramayana is one of the great Hindu epics which is about the life of Rama and his wife Sita. Rama and Sita are said to be Gods who lived in this world in human form. As per the Ramayana, Sita was kidnapped by the Demon King Raavan to the island of Lanka which is currently Sri Lanka. Rama and his brother Lakshman had to cross over a big ocean to reach Lanka and in order to help them cross the ocean bears and monkeys helped by constructing a floating bridge. It is said that the monkeys were able to build the bridge using stones that floated on the water since they wrote the name of Rama in it. This is a great epic and is believed by every Hindu. There are satellite images that show a line of limestone shoals that stretches for 29 kilometers between India and Sri Lanka which shows the possibility of the bridge which may have drowned due to the rise of sea level after the ice age. There are also many pieces of evidence that say that there are possibilities that people may have used the bridge until 4500 years ago.

There are also many other such mythological places that have got buried in Indian shores. Mahabalipuram, a port city in South India is said to have had many shore temples but only one existed. The tsunami which hit the Indian Ocean in December 2004 revealed the truth that several shore temples existed and those were submerged.

Figure 5.7: Bears and Monkeys Help Construct a Floating Bridge Between India and Sri Lanka.

Source: https://www.smithsonianmag.com/science-nature/ten-ancient-stories-and-geological-events-may-have-inspired-them-180950347/

The Lake that Exploded

The people of Cameroon were called the Kom people and they lived in Bamessi for a short time. The leader of the group was called the Fon. The Bamessi Fon planned to kill all the young men of Kom's kingdom but somehow the Kom Fon discovered this plan and wanted to take revenge. The Kom Fon called his sister and said that he would hang himself and after he died, the fluids from his body would form a lake. He advised his sister that the Kom people must not go near the lake even for fishing and leave the fish for Bamessi and leave that region. When the Bamessi entered the lake to catch fish, the lake exploded and everyone was killed.

We are not sure whether the story of Kom is true. But there are many incidents which had occurred due to the release of gases in the volcanic lakes. For example, during August 1986, a huge amount of Carbon dioxide got released from Lake Nyos which is a volcanic lake in Cameroon. This killed almost 1,700 people who were sleeping in the nearby villages. The similar gas event at Lake Monoun took the lives of 37 people. Generally, Volcanic lakes have Carbon dioxide which is dissolved at the bottom of the lake due to the pressure of the water at the top. But a sudden seismic activity can trigger the release of this gas which gives suffocation and kills people.

The tale of Kom Fon and the exploding lake might be the result of such events. People speak about stories of many deadly lakes that have killed the birds that fly above them. Such stories are famous in Africa, Greek and Rome.

The Earthshaker Named Namazu

The tale of Namazu is interesting and funny. Namazu is the name of a giant catfish which is said to have been buried under Japan. The old belief is that God Kashima has kept Namazu remain still in the land by keeping a big stone in its head. But sometimes when God Kashima slips, Namazu moves its tail or feelers which causes the surface to move.

The mythological story of Namazu is said to be the reason for the frequent earthquakes that occur in Japan. But the reality is that Japan is located at the center of several tectonic plates and with so many volcanoes and seismic activities going on, Japan by itself is a perfect place for earthquakes to happen. The story of Catfish may also be aroused due to the fact that the fish is believed to predict earthquakes. But no science or technology was able to find out the connection between the behavior of catfish and earthquakes. The only way the Japanese are detecting earthquakes is through early warning systems.

Figure 5.8: Namazu - the Giant Catfish.

Source: https://www.smithsonianmag.com/science-nature/ten-ancient-stories-and-geological-events-may-have-inspired-them-180950347/

The Chimera

The Chimera is an immortal creature and it is not human. The body of Chimera is described by Homer in the Illiad as having the front portion as that of a lion, back portion as that of a snake and the middle body as that of a goat and it snorts out flames of bright fire. As per the tales, Chimera is the daughter of the half-woman and half-snake Echidna and was killed by Bellerofonte. It is said that her flaming tongue still remains in her lair.

Figure 5.9: Yanartas, the Place Where Chimera's Eternal Flames.

Source: https://www.smithsonianmag.com/science-nature/ten-ancient-stories-and-geological-events-may-have-inspired-them-180950347/

Yanartas is a place in Turkey which is said to be the place where Chimera's tongue still sends out flames. The scientists do not have any clue about the tongue but for what they have proof is that at Yanartas, methane gas comes out of the cracks in the ground and this is the reason for it to be burning for so many years. Moreover, the flames act as a lighthouse for the sailors.

The Crater Lake - Creation

There is a great story behind the famous Crater Lake and because of this story, the Native Americans feared to even look at the lake because according to myths, anything to do with lake means to invite death. This tale was said by the Klamath people to the first Europeans when they arrived in the Northwest Pacific.

The story is that the Crater Lake was created as a result of a great battle between Llao and Skell, the rulers of the Below World and the Upper World. As the battle got fierce, darkness started to spread over the land. Lloa was positioned at Mount Mazama and Skell on Mount Shasta and they both threw rocks and flames against each other. When Mount Mazama collapsed due to the impact of the battle, Lloa was sent back to the underworld. Rain poured and filled the remaining place in the mountain thus forming a lake.

Figure 5.10: The Crater Lake at Oregon.

Source: https://www.smithsonianmag.com/science-nature/ten-ancient-stories-and-geological-events-may-have-inspired-them-180950347/

The scientists initially stated that it is not the fight between two rulers but the volcanic eruption at Mount Mazama which occurred some 7,700 years ago that had created the Crater Lake. But the scientists are now saying that the tale said by Klamath people may be true and such an incident might have happened. What made scientists come to this conclusion is the presence of red-hot rocks. There is no chance of red-hot rocks to fly through the sky when a volcano erupts and so they think that the mountain should have collapsed and filled with rainwater to create the lake.

The story of Crater Lake is very peculiar because normally mythological stories stays alive only for 600 – 700 years but this Crater story has been passed on to generations for over 7000 years now which is very rare.

The Vanished Island

The story of Teonimanu, the Vanished Island is very popular among the people of the Solomon Islands located in the South Pacific. The story goes as – a man named Rapuanate took a woman from Teonimanu to be his wife but she was taken back by her brother. Rapuanate got angry and in order to take revenge, he went to a wizard. The wizard gave him three taro plants and told to plant two on the island and to keep one with him. The wizard told Rapuanate that when the plant with him sprouts new leaves, it meant that the island was about to sink. What the wizard told happened and the island started to become salty as the water from the ocean started to rise. In order to save their lives, the people of the island fled on boats, rafts and clanged themselves to trees but the entire island got washed away.

Figure 5.11: The Vanished Island.

Source: https://www.smithsonianmag.com/science-nature/ten-ancient-stories-and-geological-events-may-have-inspired-them-180950347/

According to data facts, the island would have submerged due to some earthquake that would have made a landslide to happen. Not only this island but underwater excavations show that there are many islands that were submerged inside the water and these islands must have been inside water for a million years.

There are many such stories that are believed to have some connection with the real world. Some stories like the floating bridge of Rama are worshipped as epics and known to hundreds and thousands of people but whereas some stories like that of Teonimanu island are not known to many and not even documented in history. These are just transferred from generation to generation through word of mouth. So, when the old generation dies many of these stories may die along with them and people will not know the importance and warnings that these stories provide us.

5.5 COUNTRIES LACKING PROGRESS DUE TO RELIGIOUS RESTRICTIONS

Levels of religious restrictions and hostilities among the world's 25 most populous countries with the highest in Egypt and the lowest in Japan. The Pew Research Center found that restrictions are seen in Egypt, Indonesia, Pakistan, Russia and Turkey imposed by the society and the government.

An example is the killing of a Christian woman in March 2014 in Egypt by a group of Muslim Brotherhood supporters. A Jewish retirement home in Kansas City was targeted by a gunman in April 2014.

The highest religious restrictions imposed by the government was seen in China. Muslims were not permitted to grow long beards or wear veils during Ramadan in Kashgar Prefecture. As far as the highest hostilities in the world's most populous countries, Pakistan is high. For example, a Hindu temple set on fire after accusations or burning a Christian couple accused of blasphemy.

For Muslims, praying and prayer time is very important. "Salat" in Arabic denotes prayer time and according to Islam, every Muslim has to do Salat five times a day. During the usual workday, Salat occurs about four times. Salat is voluntary in most of the countries but Saudi Arabia is not most countries and Salat is practiced very strictly here.

From big corporate offices to small companies, every organization in Saudi Arabia must close their companies at the time of Salat so that their employees can go to the nearby mosque to pray. Salat is very important to the extent that during the 1980s, if a Muslim was seen outside during Salat, the Saudi police will themselves drag the person to the mosque.

Most of the big offices have designated prayer rooms for their staffs to pray during Salat and so the time spent by an employee for Salat in such offices is nearly 15 minutes. But in service industries such as restaurants and shops, the time given for Salat is from thirty minutes to one hour since they have to close the shop, walk to the mosque, pray, walk back and then open the shop again. For a proprietor from a foreign national, the problem would be not knowing when Salat would happen since it happens at different times each day and there is fear of the police if they don't shut down. Also, the employees who are very religious will be waiting to report any disregard to the cops.

If we analyze Salat deeply, we can understand how many working hours are being spent on it. The whole of the Saudi economy does not operate for nearly four hours every day for Salat. Apart from this, there are other normal breaks than an employee takes for having lunch, for smoking, personal works etc., all these means that on an average an employee loses 10 to 45% of his working time. When this time is turned into cash, experts say that the cost of Salat would come to about $510 billion per year and this is not a small sum of money. The global video game industry is around $99 billion per year and the Hollywood industry is about $11 billion. This makes it clear that how much amount is the Saudi government losing on Salat.

But Salat is not the only religious restriction costing Saudi's economy. Gender restriction is another major problem by which Saudi is losing more than $80 billion per year. Women employees are very less in Saudi Arabia and even the ones that were employed were paid much less than their male colleagues even though they worked harder than the men staffs. The reason why female staffs had to work so much is due to the simple fact that there were only fewer jobs for women and they have to work hard to sustain it.

Female staffs accepted less salary because going to work is the only means by which they can get out of their homes and see the outside world and be alone without their 'mahrams' (male relatives) who act as their controllers. As per the religious practice in Saudi, women cannot travel anywhere alone without their mahram or their permission.

There are also many legal procedures in hiring female staff. According to the gender segregation law followed in Saudi, the female's office must be two floors down from the male's office and that too on the other side of a digitally encoded security door. So, practically working together and completing any assignments on time will become tedious as their liberty to talk to men is also limited.

If the country does not change these norms, very soon the country will face deep financial crisis since the country is more dependent only on oil business and does not have any other varied businesses. The drop in the price of oil exports is a great warning signal.

A country's economy grows as its labor force grows. According to the World Bank, the female labor participation in Saudi is around 21% only which is very low. Saudi Arabia will be able to increase its economy from somewhere between $80 billion and $100 billion if it reduces the gaps in female employment and removes some restrictions. Many foreign consultants and diplomats are continuously pushing the Saudi government to bring about reforms to increase the rate of female employment and thereby increase its economy. 'Vision 2030' plan seems to be an initial step in the process of bringing about 30% female employment.

The country may bring in any new plan but unless the religious restrictions are relaxed or removed, there will be no room for improvement, no matter how much training is given to female staffs. Women represent half of the workforce and when this half workforce sits at home just because they are not allowed to talk to men, not allowed to drive or not allowed to go to any government ministries, it does not reflect a healthy nation or an economy.

REFERENCES

1. http://www.yourarticlelibrary.com/religion/essay-on-religion-meaning-nature-role-and-other-details-5931-words/8586
2. https://www.sharecare.com/health/spiritual-wellness-religion/rituals-and-spirituality
3. https://www.scienceforums.net/topic/71971-the-importance-of-ritual-in-religion/
4. https://www.sbs.com.au/topics/life/culture/article/2016/06/27/why-rituals-are-still-relevant
5. https://en.m.wikipedia.org/wiki/Ritual
6. https://www.quora.com/Why-do-religions-have-rituals
7. https://www.rookiemag.com/2012/12/of-ritual-importance/
8. http://www.studyreligion.org/concerns/living.html
9. http://www.yourarticlelibrary.com/religion/essay-on-religion-meaning-nature-role-and-other-details-5931-words/8586
10. https://journals.sagepub.com/doi/pdf/10.1177/1524500416672439
11. http://thinkbusiness.nus.edu/article/the-impact-of-superstition-writing-on-the-wall/
12. http://www.claudiageib.com/writing/2017/9/25/how-superstitions-and-myths-affect-animal-conservation
13. https://www.atlasobscura.com/articles/endangered-animals-cultural-myths
14. http://article.sapub.org/10.5923.j.ijpbs.20180804.02.html
15. https://countrynavigator.com/blog/cultural-intelligence/superstition/
16. https://www.popsci.com/science-superstition#page-5
17. https://en.wikipedia.org/wiki/Sacrifice
18. https://friendlyatheist.patheos.com/2016/03/15/how-hindu-religious-rituals-harm-the-environment/
19. https://www.smithsonianmag.com/science-nature/ten-ancient-stories-and-geological-events-may-have-inspired-them-180950347/
20. http://www.pewresearch.org/fact-tank/2016/06/28/religious-restrictions-among-the-worlds-most-populous-countries-2/
21. https://www.vox.com/world/2017/1/17/14257816/saudi-arabia-economy-islam-religion

Importance of Environment

6

"Cherish the natural world because you are part of it and you depend on it."

Sir David Attenborough

CONTENTS

The environment is very vital for humanity to survive. We all rely upon environment more than anything else in this world because it is the treasury from which we get everything. Basic needs such as water, food, clothing, shelter, etc., are obtained from the environment. Not only food, but the environment also provides us with different minerals and metals. It is the environment that provides some of the best medicine in the world. The environment also acts as the greatest source for entertainment. This Chapter throws more light on all these aspects in detail.

6.1 ENVIRONMENT IS THE NATURAL RESOURCE FOR FOOD

There is no question that Earth has been a giving planet. Food, water, medicine, materials for shelter, climate and the natural cycles are all gifts from nature to humans . These are referred to as by scientists, yet an example is the caves paintings at Lascaux. While nature provides food such as foods specific for a region; the development of trade between nations have allowed for the availability of several tropical fruits across the globe.

All before the arrival of the human, the salt water was separated from the fresh water, dry land was made, gardens were planted with fruit-bearing trees and animals and fish were made. The planet already had what we needed before we were created!

Nature provides us with everything we need and the less we interfere with it, the better it is for us! Check out these amazing 'Nature's Pharmacy' items that are available to us! A sliced carrot looks just like a human eye. You can see the pupil, the iris and even the radiating lines. And, YES, modern science has shown that the nutrients in carrots greatly contribute to the health of the eye.

rapes in a cluster looking like the shape of the heart. With research showing the functioning of grapes as foods good for the heart . A walnut looks like a little brain, a left and right hemisphere with upper and lower cerebellums. Even the 'wrinkles' on the nut are just like the neo-cortex. We now know walnuts help in the formation of more than three dozen neuron-transmitters for brain function.

Kidney beans actually help heal the kidney and maintain it's function. And yes! They look exactly like the human kidneys! Celery and bok choy look like bones to boost bone strength as they contain the same amount of sodium in bones that is withdrawn from the bones if the levels dip in the body.

There is no physical substance people require more than freshwater: without water, we can just endure a couple of shocking days. While contamination and abuse have compromised huge numbers of the world's drinking water sources, nature has an out-dated arrangement, in any event, to contamination. Sound freshwater environments—watersheds, wetlands, and woods—normally clean contamination and poisons from water. Soils, microorganisms, and plant roots all assume a job in sifting and reusing through poisons with a cost far less expensive than structure a water filtration plant. As per investigate, the more biodiverse the biological system, the quicker and all the more proficiently water is refined.

It is difficult to accept that there is a lack of clean water on the planet when 70 % of the surface is secured with water. Fortunately, nature has a method for decontaminating contaminated water through its freshwater biological systems, where the dirt, microbes and plants channel out poisons from water, just like the case with common wetlands and woodland waters. There are still spotless water streams and springs, wells and waterways from which water can be consumed, however sadly their number is diminishing and we have to recollect exactly how basic new water is to us and make a move to clean the world waters.

Fisheries:

Water has been a source of food since history. Today, in the midst of worry of a worldwide fishery breakdown, in excess of a billion people rely upon fish as their essential wellspring of protein, a considerable lot of them among the worldwide poor. Fisheries additionally give employment, both straightforwardly and in a roundabout way, for around a large portion of a billion. Coral reefs, mangroves, and seagrass biological systems give nurseries to the world's fisheries, while the untamed sea is utilized for moving courses and hunting.

Indeed, even with the immediate significance of the world's fisheries for nourishment, stewardship has been missing, enabling numerous populaces to drop sharply and as yet allowing biologically dangerous angling. While the world's fisheries are essentially undermined by overfishing, including by catch, marine contamination is likewise a noteworthy issue.

The sea covers around 70% of the Earth's surface, which means it is the biggest condition for living things on Earth. There are five noteworthy sea natural surroundings: tropical or reef living space, calm waters, vast sea, the remote ocean, and polar districts. Most ocean life lives in the best 150 meters (500 feet) of the sea. From the warm, sunlit waters of coral reefs to the dim, cold waters of the remote ocean, the sea abounds with life.

The sea is a significant wellspring of sustenance and different assets. Since a long time before written history, people have utilized the ocean as a wellspring of nourishment. While just 5% of the protein devoured by world populaces originates from the ocean, it is as yet a significant commitment to the eating routine of a large number of the world's occupants

6.2 ENVIRONMENT IS A TREASURY OF ENERGY RESOURCE

Solar is the most abundant energy resource on planet Earth. Even after accounting for weather variation, the average solar power received by the continents alone peaks at 23 million gigawatts thus dwarfing all other wind, hydropower and geothermal sources. Despite the availability in countries with little cloud, it has not been developed as:

- The space requirements.
- Solar energy is costly.

Intermittency caused by weather, day-night cycles and seasons is a showstopper

Natural resources are materials and substances that happen normally and can be utilized for financial increase. They incorporate minerals, timberlands, land, and water. Some common assets, for example, soil and water, are basic for the presence of life.

A non-renewable natural resource is characterized as an asset that can't be supplanted in our lifetime. They incorporate metal minerals, petroleum derivatives, earth minerals, and in some in specific circumstances groundwater. Inexhaustible normal assets are assets that can be supplanted normally in our lifetime, and utilized over and again. Models incorporate crisp water, timber, oxygen, and sunlight based vitality.

This segment will talk about the top natural resources and what they are utilized for.

Copper

Copper is said to be one of the principal metals utilized by man thus it is being used for about a large number of years. Copper is acquired normally. Copper is helpful; it is utilized in electrical wires, businesses, hardware, material and pipes. Copper is likewise utilized in nourishing enhancements and fungicides for horticulture.

The United States is the second biggest maker of copper, which is mined from copper minerals and recouped through reusing. Copper is anything but difficult to get as it very well may be reused absent much misfortune in quality. Thus the measure of copper acquired from the mines can be gotten through reusing process.

Helium

The vast majority of us knew the utilization of Helium as a filler in inflatables; however, inflatables are only a little division of its whole use. Helium has other wide uses as a superconducting magnet in MRI scanners and NMR spectrometers, cooling for the Large Hadron Collider (LHC), and satellite instruments.

Since helium is non-responsive it is utilized to make a protective inert atmosphere in the procedures of making fiber optics and semiconductors. Helium is likewise used to distinguish for breaks in cooling frameworks, and to blow up air-packs in vehicles since it is non-harmful and diffuses rapidly.

Despite the fact that helium is the second most inexhaustible component in the close planetary system, it is considered a non-sustainable resource. Most helium is delivered where fortuitous events of far-fetched circumstances happen inside the Earth's outside, making the gas uncommon. Removing helium from the air is uneconomical with there being under 0.00052% in air.

Bauxite

Bauxite is present as rocks in soils with next to non-dissolvable materials that are situated in wet tropical atmospheres. This is one reason why Australia is the top maker of bauxite. Bauxite is the fundamental material in making aluminum metal, and about all aluminum delivered has been separated from bauxite. It is likewise utilized in pressure driven breaking, a procedure in boring for oil and gas.

Iron

Everybody knows iron since we use it consistently and in numerous items and it is likewise the world's most regularly utilized metal. Iron is a decent conductor, tough, and ample making up 5% of the Earth's crust. This is the reason it is utilized to make steel, electrical wiring and direct, funnels, instruments, pivots, entryway handles, stoves, pots and skillet, and even couch outlines. The vast majority of the world's iron mineral is mined from united iron developments (BIFs) basically situated in Australia, Brazil, and South Africa.

Coal

Individuals have been utilizing coal as a warmth source for quite a long time. In the present society, coal has turned into the essential wellspring of fuel for power creation since it is plenteous, has high vitality yield, and is moderately cheap. Coal is a non-renewable energy source made through a procedure known as coalification. Plants that shaped a great many years back in bog timberlands passed on and framed layers of peat that were covered through geological processes, and afterward modified through warmth and weight in low oxygen conditions.

There are numerous issues that go with the mining and utilization of coal. As per the U.S. Geographical Survey, one of the more genuine ecological issues is the unfavorable impact on surface and groundwater from the transfer of powder and slime that outcomes from the consuming of coal and the cleaning of vent gases.

Petroleum gas

It is utilized as a vitality hotspot for warming, cooking, and age of power. It is additionally utilized during the time spent making plastics and other financially significant natural mixes.

Petroleum gas is a non-renewable energy source made by the decay of natural material, as a rule from antiquated marine living beings, in an anaerobic situation. There are two classes of flammable gas stores, conventional and unconventional. Ordinary ones are related to oil stores, and the latter stores incorporate coal bed methane, shale gas, and tight-gas sandstone.

Oil

Like petroleum gas, oil is made through the disintegration of the natural issue in an anaerobic domain more than a huge number of years. Oil supplies 40% of the country's vitality needs and is utilized for the generation of gas, diesel fuel, fly fuel, propane, and black-top. Oil is additionally utilized in the assembling of petrochemicals for creation of plastics, engineered elastic, and synthetics.

The area of oil and gas stores is controlled by plate tectonics. This is the reason most oil is found in deserts, arctic locales, waterway deltas, and mainland edges seaward. Plate tectonics make the areas for anoxic burial and the high weight and warmth required for the natural issue to go to oil.

Salt

Salt, otherwise called sodium chloride, is a fundamental mineral for the life of people and creatures. Salt is mined through one of four strategies: rock salt mining underground, solution mining by infusing a solvent that breaks up underground salt and afterward is recuperated through sun based drying, sea water evaporation in which ocean water is gathered and evaporated.

Over 40% of the salt mined is utilized in the chemical business, with another 40% being utilized as a deicer on streets in the winter. The staying 20% is spread out over preparing of table salt and assembling of elastic and different merchandise.

Timber

Timber is a significant inexhaustible normal asset in numerous social orders the world over. On the off chance that generation is drawn closer in a mindful manner, it very well may be both maintainable and prudent. Timbers are for the most part used to make amble for use in development. They can likewise be generally formed and utilized as substantial bars and segments. Timber is additionally used to make paper items, fiber board, hard board, pressed wood, and molecule board.

Soil

Soil is a standout amongst the most significant regular assets on Earth, being required both straightforwardly and by implication for sustenance creation, assembling of mechanical crude materials, and energy sources. Soil is fundamental for the capacity of biological systems giving supplements, oxygen, water, and warmth.

Soil resources are being corrupted by poor rural practices and chemical contamination. A standout amongst the most noteworthy difficulties confronting present and future ages is the conservation of this essential common asset from contamination and physical loss.

Non-sustainable power resources

These resources have shaped more than a huge number of long stretches of geographical procedures, and we're utilizing them quicker than they can be normally renewed. Models incorporate non-renewable energy sources like coal, oil, and gaseous petrol. We as of now rely upon non-renewables to meet a large portion of our vitality requests, separating and combusting them essentially to produce power or create energizes for transportation.

While we frequently think about atomic power as an elective vitality alternative, it is critical to recollect that while its carbon yield is low contrasted with petroleum products, atomic splitting is as yet a non-inexhaustible asset.

Figure 6.1: Bituminous Coal.

Source: https://www.e-education.psu.edu/geog432/node/114

Coal

A combustible sedimentary rock made up mostly of carbon and hydrocarbon, coal is the most abundantly used fossil fuel worldwide for the generation of electricity. In the United States, around 93% of the coal expended is utilized to create power (EIA site). The steel, cement, and paper enterprises likewise depend intensely on coal for both warmth and results. The ignition of coal results in right around 3 fold the number of CO2 discharges as the measure of coal combusted - copy a pound of coal, radiates very nearly 3 pounds of carbon dioxide into the air

Natural Gas

Figure 6.2: A natural gas flame.

Source: https://www.e-education.psu.edu/geog432/node/114

Natural gas is a gas that is made up mostly of methane and found near other fossil fuels. Methanogenic activity happening in landfills and bogs additionally produces this gas. Like oil, natural gas should initially be prepared before we can utilize it as a fuel. It is critical to evacuate the greater part of different segments of flammable gas until it is absolutely methane. Whenever combusted, natural gas creates just about a large portion of the ozone-depleting substance outflows as coal does, making it a prominent petroleum product in a carbon-conscious society.

Petroleum

Figure 6.3: Global Sante Fe Rig 140, owned by Transocean.

Source: https://www.e-education.psu.edu/geog432/node/114

Petroleum is a toxic, flammable liquid occurring in geologic formations beneath the earth's surface (also known as crude oil). We use oil for a wide assortment of things - the biggest utilization of oil is for fuel oil and gas. Be that as it may, you'll discover oil in spots you probably won't anticipate it too - pharmaceuticals, plastics, black-top, lamp fuel, and manufactured elastic, to give some examples. Like flammable gas, the oil must be handled before we can use it. Raw petroleum normally contains a wide range of sorts of hydrocarbons.. So, to process the oil for a specific application, the crude must be heated to a specific temperature range.

Uranium

Figure 6.4: A Nuclear power plant.

Source: https://www.e-education.psu.edu/geog432/node/114

Uranium is a very heavy, fissile metal (U-235) that can be split in a fission chain reaction, to form large amounts of heat to generate electricity. Since one response sets off another 2-3 responses, the subsequent chain response is exponential and enables us to create a generous measure of heat with a moderately modest quantity of uranium. After extraction, uranium must be handled before we can begin the splitting procedure. Most reactors use uranium that has been finely ground and after that gasified to uranium hexafluoride which is then changed over to pellets of uranium dioxide. While atomic power offers us greenhouse gas emission-free source of energy, the worries related to reactor disappointment and long haul stockpiling of spent fuel present huge obstructions to far-reaching appropriation.

Renewable energy resources

Renewable energy resources can be replenished hence, not subject to depletion when used. The examples are summarized below:

Fossil fuels have been used despite the benefits of the above due to:

- inexpensive
- in abundant supply
- storable

Wind

We can give wind something to do by using it to control turbines that produce power. The cutting edges of the turbines turn a pole which at that point controls a progression of riggings that feed into a generator and produce power. Wind turbines can be arranged on horticultural or forested land, so there's practically no compelling reason to clear new zones of land to site wind fields. A few people disagree with the clamor and visual contamination of the breeze turbines. There are likewise worries about bird and bat mortality..

Solar

Figure 6.5: Solar Panels on a sunny day.

Source: https://www.e-education.psu.edu/geog432/node/114

Geothermal

Figure 6.6: A geothermal plant.

Source: https://www.e-education.psu.edu/geog432/node/114

The center of the Earth produces heat, and in numerous spots, the world over we can tap that warmth to the surface to warm and cool homes. Like sun based and wind, geothermal power can be caught at an individual level - you introduce a geothermal warmth siphon to cool and warmth your home - or at an increasingly brought together scale by utilizing dry steam or boiling water to create power. Since no fuel is combusted, the plants discharge no ozone-depleting substance like non-renewable energy sources. There are minor discharges causing acid rain.

Biomass

Figure 6.7: Biomass from fuels reduction on private lands in northeast Oregon.

Source: https://www.e-education.psu.edu/geog432/node/114

Biomass energy envelops a wide scope of energizes including wood and other plant material, sustenance waste, trash, and excrement. On account of wood and plant material, it very well may be prepared and combusted, for example, producing power or a kettle. With sustenance waste and fertilizer, the gases coming about because of the procedures of decay of these materials can be saddled and combusted for useful end use. Using biomass for vitality not just furnishes us with a sustainable power source however regularly enables us to put a material that would somehow or another be viewed as waste to decent use. It's critical to recall that while a sustainable power source, biomass fills each has its very own arrangement of ecological effects to consider..

Hydropower

Figure 6.8: A hydropower plant.

Source: httprs://www.e-education.psu.edu/geog432/node/114

Water is put away in a repository, producing a ton of potential energy. At that point it is constrained through a dam, turning a turbine, which at that point turns a generator to create power. Utilized water is then come back to the waterway. While much work has been done to guarantee that hydroelectric power has an insignificant negative effect on nature (counting the capacity of hydropower offices to procure a low-sway rating), it is critical to perceive the biological system disturbances this vitality asset can cause. Damming a waterway and using the water to create power adjusts the oceanic and riparian biological systems, can restrain the capacity of fish to explore over their normal natural surroundings degree and can change the temperature and synthesis of the stream itself.

6.3 ENVIRONMENT – THE BEST SOURCE OF RECREATION

The involvement of people in volunteering or recreation can allow for natural resources.

History has proven that the environment of a place leads to the existence and development of tourism. Outdoor recreation lovers are prone to choose places with eye-catching sceneries, beautiful weather and special landscape characteristics as their outdoor activity locations. According to Fishbein & Ajzen (1980) and Knapp (1999), attitude towards the environment is the main factor that affects human behavior in the environment. Meanwhile, Wright (1991) said that recreation is a free time activity for the purpose of leisure including after work activities. The exhilaration of outdoor recreation activities has been a trend in today's world. Generally, this would cause direct and indirect impacts on nature and all living organisms. Thus, all activities need to be practiced and done based on all environmental codes and ethics. This is important for the next generation to appreciate the treasures of nature more than today's generation. Based on this phenomenon, the writer will unravel and discuss the issues and challenges of outdoor recreation towards the environment.

Expanding urbanization and the medical advantages of a functioning way of life have made an interest for outside diversion. Timberlands give a domain where it is conceivable to escape from the worries of present-day life to interface with nature and the wide open and experience serene and picturesque spots. Woodlands additionally give wild territories where exercises, for example, mountain biking, and other dynamic and testing sports can be sought after. The advancement of woods for amusement is viewed as a significant part of feasible ranger service and there is an expanding acknowledgment of the wide going advantages that timberlands give to society. Today, diversion is maybe the most significant non-timber administration given by woodlands.

Since the mid-1970s there has been a functioning system of giving recreational offices in Irish state woods. Until this point in time, somewhere in the range of 150 recreational locales and 11 wood parks have been built up in backwoods all through the nation. The vast majority of these offices were set up by the Forest Service which today is being kept up and extended by Coillte. Notwithstanding the arrangement of recreational destinations – outing territories, preliminaries and so forth. Coillte additionally has an

Open Forest Policy that enables the free community to all pieces of its 445,000ha timberland domain.

Backwoods are a truly important entertainment asset drawing in a large number of guests every year (evaluated at 18 million visits to Coillte timberlands by Fitzpatrick Associates in 2005). Woods as the travel industry framework contribute altogether to the financial yield of the nation. The open great advantages are additionally very important in supporting the woods business by expanding the arrival on speculation and picking up help for improved interest in backwoods related exercises.

Open air diversion is the most ideal approach to connect individuals and nature. Be that as it may, if individuals ignore the principles and guidelines on saving nature, they will be accused by the future age for harming the whole biodiversity.

Family bonds are improved by the sharing of relaxation time. Families that reproduce together will in general be nearer, progressively strong and improve their odds of remaining together.

6.4 ENVIRONMENT – THE BEST PHARMACY IN THE WORLD

People have an inborn association with nature. Research demonstrates that simply being around greenery can help mindset and lower circulatory strain, just as improve consideration and the capacity to ponder an issue.

Nature is our most prominent prescription bureau: to date, it has given mankind a huge number of life-sparing medications from quinine to headache medicine and from morphine to various malignant growth and HIV-battling drugs. There is no doubt that furthermore significant meds—maybe even marvel fixes—lie undiscovered on the planet's biological systems. Truth be told, specialists gauge that under 1% of the world's realized species have been completely analyzed for their therapeutic worth. Nonetheless, the environments that have yielded a portion of the world's most significant and promising medications, for example, rainforests, peat bogs, and coral reefs—are additionally among the most imperiled. Safeguarding biological systems and species today may profit, or even spare, a large number of lives tomorrow.

Ongoing exploration has discovered what nature lovers have since quite a while ago anticipated: investing energy in a green space, for example, a recreation center gives advantages to one's psychological and physical wellbeing. Practicing in a recreation center, rather than inside a rec center, has appeared to give psychological well-being benefits as a more prominent feeling of prosperity. Strolling for 20 minutes in a green space has been demonstrated to assist kids with ADHD improve their fixation, notwithstanding filling in also, or better, than medicine. Individuals who live in progressively regular settings have better by and large wellbeing, notwithstanding when research has considered monetary contrasts.

Herbs, plants and seeds are normal drugs we can utilize – as oils, teas, powder, concentrates, substances, etc., and they are as yet the most productive in mending.

Fifty years back, the World Health Organization (WHO) characterized wellbeing as "a condition of complete physical, mental and social prosperity and not only the nonappearance of malady or ailment." This definition should fill in as an update that rethinking the perspective on ecological wellbeing and the indigenous habitat requires numerous movements in deduction, just as a readiness to seek after a decent variety of methodologies.

What's more, contact with the natural world might be legitimately helpful to wellbeing. Assuming this is the case, at that point the field of ecological wellbeing needs to stretch out past insignificant contemplations of lethality. As indicated by the speakers, research and arrangement must be coordinated toward understanding the communications of human wellbeing and the regular habitat, from the closest to home level—that is, the manner by which people associate with their general surroundings—to the worldwide scale—that is, the way countries consent to diminish contamination, seek after sound horticultural and land use practices, and feed their residents.

A hypothetical reason for the thought that contact with nature is helpful originates from E.O. Wilson, who presented the term Biophilia very nearly 20 years back, characterized as the inherently passionate association of people to other living beings. From a developmental point of view, a profound situated association with the natural world ought to be nothing unexpected. People have been advancing for in excess of 2 million years, yet have lived moderately protected from nature for just the most recent 10,000 years.. As Wilson (1993) noted,

"It would be quite extraordinary to find that all learning rules related to the natural world have been erased in a few thousand years, even in the tiny minority of peoples who have existed for more than one or two generations in wholly urban environments."

This is certainly not another thought, guaranteed Howard Frumkin of Emory University. The human association with nature and the possibility this may be a segment of good wellbeing have a long and interesting history in reasoning, craftsmanship, and mainstream culture. There is abundant proof—from creatures, plants, scenes, and wild encounters—that we can expand on this alliance to upgrade our wellbeing.

Animals have constantly had a noticeable impact on human life. Today, a larger number of individuals go to zoos every year than to all games. The greater part of all U.S. families owns pets. Creatures contain in excess of 90 percent of the characters utilized in language securing and including in kids' preschool books. An extensive assemblage of proof connections contact with creatures to human wellbeing. As indicated by Frumkin, saving the security among individuals and their creatures, such as empowering great sustenance and exercise, seems, by all accounts, to be to the greatest advantage of those worried about general wellbeing.

Introduction to plants and flowers additionally sustains a feeling of our prosperity. In a 1998 National Gardening Survey of in excess of 2,000 arbitrarily chosen family units, half of the respondents concurred with the explanation that blossoms and plants at amusement parks, noteworthy locales, fairways, and eateries are essential to satisfaction in these areas, and 40 percent concurred with the explanation that being around plants makes them feel quieter and progressively loose.

Office representatives report comparable sentiments, expressing that an office with plants is an increasingly alluring work environment. Analyst Michael Perlman has composed of the mental intensity of trees, as to prove by folklore, dreams, and self-announced passionate reactions. For sure, the idea that plants have a job in psychological wellness is genuinely entrenched, said Frumkin. Of note, medical clinics have generally had gardens as an aide to recovery and mending. Maybe this long-established practice mirrors an acknowledgment that vicinity to plants, similar to closeness to animals, may in certain conditions improve wellbeing.

To come back to a developmental point of view, mankind's history most likely started on the African savanna, a district of open meadows punctuated by dispersed stands of trees and denser woods close streams and lakes. At the point when offered a selection of scenes, individuals respond most decidedly to savanna-like settings, with moderate to high profundity or transparency, generally smooth or uniform-length green vegetation or ground surfaces, dissipated trees or little groupings of trees, and water, a finding that is steady over each culture considered.

This impact may stretch out past feel, to reclamation of wellbeing or recuperation from stress. Research on recreational exercises has demonstrated that individuals seeing savanna-like settings report sentiments of quietness, serenity, or unwinding. On psychometric testing, individuals survey such settings show diminished dread and outrage, upgraded positive effect, and improved mental sharpness, consideration, and intelligent execution.

The same results emerge from studies that directly consider conventional health end points. In 1981, University of Michigan architect Ernest Moore found that "prisoners assigned at random to cells along the prison's outside wall with a view of rolling landscape and trees, had significantly lower frequency of sick call visits compared to those in the inside cells." Moore could not identify any design features to explain this difference and that the outside view may provide some stress reduction.

S A 1984 paper in Science titled "View Through a Window May Influence Recovery from Surgery". "Postoperative patients, assigned essentially randomly to rooms with views of trees, had statistically significantly shorter hospitalizations, less need for pain medications postoperatively, and fewer negative comments in the nurses' notes, compared to patients with the brick views" (Ulrich, 1984).

Physician David Cumes has described "wilderness rapture," including self-awareness; feelings of awe, wonder, and humility; a sense of comfort in and connection to nature; increased appreciation of others; and a feeling of renewal and vigor. These outcomes are often cited in favorable accounts of so-called wilderness therapy for a variety of psychiatric, organic, and emotional disorders. Most published studies related to mental health end points, and although they are limited, many do suggest some benefit from wilderness experiences. Mental health has been studied more than somatic conditions, and short-term benefit has been demonstrated more than long-term benefit " (Cumes, 1998).

Biodiversity and Human Health

There is a powerful connection between the biodiversity and our own well-being evident from medicines from nature, and as indicators of pollution-related disease.

In the pharmacopeia, the natural world is constantly offering exacerbates that are helpful in prescription, and the quest for organically dynamic mixes proceeds. Tom Lovejoy of the World Bank gave instances of this, including mixes acquired from ooze molds from the banks of the Zambezi River that are viable in the treatment of tumors impervious to taxol; the immunosuppressant cyclosporin, which originates from a parasite; analgesics from tropical ocean creatures and toxin frogs; and lovastatin, which originates from the bacterial world.

"Biodiversity is essential for guaranteeing the healthiness of humans." —Tom Lovejoy

Characteristic items are frequently basic in both research and analysis, said Lovejoy, the prime model being the polymerase chain response, which depends on a natural substance to start the response. Another model includes horseshoe crabs, the blood of which is helpful in recognizing certain impurities in medications. Different species can likewise give sudden understanding into our own science; for instance, a pit snake from South America has a venom that causes its unfortunate casualties' pressure to drop to zero for all time. Brazilian researchers concentrating this snake 20 to 30 years prior revealed the angiotensin arrangement of pulse guideline. This exploration, in the end, prompted the advancement of a class of medications known as ACE (angiotensin-changing over protein) inhibitors.

"Biological diversity also serves as a set of indicators of environmental change, for example, the study of endocrine-disrupting chemicals in wild species and the decline and disappearance of the world's amphibian species. Biological diversity becomes the ultimate integrator of all the things affecting living systems", said Lovejoy. An extra manner by which biodiversity interfaces with human wellbeing is in diminishing environmental infection. There is the use of organisms, to tidy up lethal waste issues in the earth, especially PCBs, oil slicks, and substantial metals. Organic frameworks progressively are being viewed as potential methods for tidying up the water supply through the rebuilding of watersheds.

Good judgment reveals to us that a sound situation is imperative to human wellbeing. Be that as it may, just as of late have science and innovation furnished us with approaches to gauge the relationship between a sound body and environment. The regular habitat wherein we go through our days and the national and worldwide network wherein environmental securities must be arranged give both a local and a worldwide point of view by which to think about ecological wellbeing.

"Although there clearly is interdependence between public health and the environment, workshop participants noted that we have limited resources for identifying and understanding challenges to health or implementing intervention strategies. Some of the higher-order issues, such as sustainability, must be addressed if we are to achieve better health" , noted Rafe Pomerance.

"Another central quandary is the reduction and disposition of waste. We cannot continue to have consumption that outweighs the production capacity of our ecosystems, and we cannot continue to produce waste at a rate that outweighs our ability to assimilate it back into the ecosystems without negative impacts", said William McDonough.

A second challenge is development of baseline data on different environmental stressors. Effects of poor public wellbeing can be immediate or roundabout. Speakers noticed that we have tended truly to concentrate on the immediate impacts of contamination on general wellbeing—for instance, lethality or adverse wellbeing impacts—and less on the biomarkers that can gauge immediate and circuitous impacts through effects on environmental frameworks.

To address these difficulties, a few workshop members expressed that we have to grow progressively comprehensive and coordinated ways to deal with natural wellbeing that fuse contemplations of human organic and environmental wellbeing and accomplish better comprehension of these interrelationships. The frameworks perspective on the biologist must be received as another worldview for the ecological wellbeing researcher.

6.5 SUMMARY

The ecosystem (every one of the networks of living beings found in a particular spot, their environments, and their collaborations) where we live gives services to people and every single other species that are basic to our wellbeing, personal satisfaction and survival. For instance, trees expel carbon dioxide and different toxins from the air we inhale and furthermore cool our air temperatures, diminishing the arrangement of ground-level ozone, a contamination that can make heart and lung issues intensify; our wetlands store tempest water, channel and make innocuous tempest water poisons, and revive our aquifers (where the greater part of us get our drinking water) with these separated waters; and the dunes on our shorelines structure normal obstructions to waves and give significant territory and travel ways for wildlife.

The following are three natural objectives that are basic to accomplishing manageable social orders and being great ecological stewards, a portion of their financial, social and ecological advantages, and a portion of the moves we can make to accomplish these objectives.

Here are such a large number of simple approaches to turn out to be progressively feasible. We are generally animals of propensity, and changing our conduct takes some time. Little advances are critical, and once you discover how simple each is, you will be amazed how rapidly you can have a beneficial outcome in securing our condition for your family and your planet.

The environment is the sole factor for the presence of life on earth. Without it, there can be no life on earth. In our nearby planetary group, there are other enormous planets, be that as it may, have no life because of the absence of the conditions.

Subsequently, even researchers are taking a shot at just Mars to make life on it as it has some appropriate condition. This demonstrates the significance of the condition forever and presence.

So underneath we will see the significance of condition for different jobs throughout everyday life.

- For good living: To have a sound life one needs outside air (free of contamination), clean water supply and clean environment. In the event that air and water are contaminated and surroundings are dirty, at that point we are inclined to lethal wellbeing issue. So we have to protect conditions as far as contamination and transfer of trash.
- Better breath: All living things inhale air for survival. The air for the most part makes out of nitrogen, oxygen, carbon-dioxide and different gases. Of them oxygen is the key for the body as it helps in age of vitality from the sustenance overwhelmed by the assistance of mitochondria.
- Better water: Water includes almost 80% of all-out weight. It is the mechanism for all the response and physiological procedures in the body. On the off chance that the very water is sullied by foulness or poisons or even malady causing microorganisms like microbes, at that point that water is an incredible hazard to drink.
- Better precipitation: Rainfall is one of the means of the water cycle on the earth. Without rain, there will be no water on the earth (other than ocean water which is unfit to drink). This rain relies upon trees and green vegetation around. The nearness of plants,

trees and greenery is fundamental in the earth for precipitation, natural air and water.

- Better soil: Soil is different segments of condition. Soil is valuable for the development of plants which are again a wellspring of sustenance to the creatures around. This top layer is useful for plant development and seed germination. Further, it is an extraordinary normal asset for some metals and chemicals and so on. So we have to ration soil for better condition.
- Better plants: Plants and trees are likewise key in the significance of condition. Plants which incorporate trees, bushes, hedges and even little grass help in refining air as they produce oxygen and take up carbon dioxide. Consequently, they keep oxygen levels noticeable all around.

Significance of Environment for people

- For sustenance: Food materials are acquired from plants, creatures and normal sources. We acquire sugars, fats, proteins from plants. From animals, we get meat (protein), milk, eggs (fowls), nectar (honey bees).
- For shelter: The required structure material including furniture is acquired from nature around.
- For apparel: The attire necessities like cotton strands, wool, and jute is given from plants and animals.
- For furniture: Though we have plastic and misleadingly made furnishings, nothing can supplant furniture from wood, porcelain, china and so forth.
- For medicines: We get numerous medicines from nature like from plants, animals (cod liver oil) and so forth. These prescriptions are good to people with negligible or no symptoms than artificial medications.
- Significance for Health.

No stress: An individual in sitting in an office or shut room is inclined to more worry than the one working in open condition like ranches, seashores and so forth. Henceforth we see numerous colleges and research foundations are worked in the midst of condition far from the city.

Happy living: To the individuals who wish to carry on with a wonderful life, it is smarter to fabricate a home in spots of the seashore or in the middle of farms and so on. One encounters genuine feelings of serenity, low aggravation in mindset in such places.

Forestall spread of infections: There is particularly chance for the spread of sicknesses in shut conditions where there are numerous individuals bound to little space.

Considering the significance of condition with reference to above focuses everybody needs to avoid potential risk to limit squander via cautious utilization of energy, vehicles and other contamination producing operators.

REFERENCES

1. https://news.mongabay.com/2011/04/what-does-nature-give-us-a-special-earth-day-article/
2. https://ocean.tamu.edu/academics/resources/ocean-world/fisheries/the-ocean-environment-as-a-food-resource/index.html
3. https://highstylife.com/great-things-nature-gives-us/
4. https://www.collagevideo.com/blogs/healthy-tips-recipes-by-gilad/52205252-nature-provides-us-with-everything-we-need
5. https://www.scientificamerican.com/article/a-way-to-the-most-abundant-energy/
6. https://www.renewableresourcescoalition.org/top-natural-resources/
7. https://www.e-education.psu.edu/geog432/node/114
8. https://www.forestryfocus.ie/social-environmental-aspects/recreation/
9. http://www.academia.edu/2905352/Impact_of_Outdoor_Recreation_On_Environment
10. https://www.environment.gov.scot/our-environment/people-and-the-environment/recreation/
11. https://www.ncbi.nlm.nih.gov/books/NBK99584/
12. https://www.hiltonheadislandsc.gov/sustainability/whycare.cfm
13. https://www.studyread.com/importance-of-environment-conservation/

Managing the Environment

7

"The Earth does not belong to us: we belong to the Earth."

Marlee Matlin

CONTENTS

We have seen the importance of the environment and its uses to mankind as well as the entire world in the earlier chapters. This makes the environment very significant and we should learn to protect and preserve our environment. This Chapter explains what managing environment means and it highlights the various means and ways by which we can save our environment. It also explains the importance of preserving our natural resources and why the environment needs to be protected.

7.1 MANAGING NATURAL ENVIRONMENT – DEFINITION AND OVERVIEW

The natural environment encompasses all living and non-living things occurring naturally, meaning in this case, not artificial applied to the earth. This environment encompasses the interaction of all living species, climate, weather and natural resources that affect human survival and economic activity. The concept of the natural environment includes the following components:

Complete ecological units that capacity as regular frameworks without huge humanized human intercession, including all vegetation, microorganisms, soil, rocks, climate that happen inside their limits and their inclination.

Individuals only sometimes find completely indigenous habitats on Earth, and instinctive nature more often than not shifts in a continuum. All the more exactly, we can think about the various angles or parts of the environment, and see that their level of instinctive nature isn't uniform. On the off chance that, for example, in a farming field, the mineralogic piece and the structure of its soil are like those of an undisturbed woodland soil, yet the structure is very extraordinary.

Indigenous habitat is frequently utilized as an equivalent word for living space. For example, when we state that the regular habitat of giraffes is the savanna.

Environments

The importance of the interrelated set of relationships is established in an ecosystem.

CEugene Odum, one of the founders of the science of ecology, stated: "Any unit that includes all of the organisms (i.e.: the "community") in a given area interacting with the physical environment so that a flow of energy leads to clearly defined trophic structure, biotic diversity, and material cycles (i.e.: exchange of materials between living and nonliving parts) within the system is an ecosystem."

A more prominent number or an assortment of species of a biological system may add to more noteworthy resilience of an environment, in light of the fact that there are more species present at an area to react to change and hence "retain" or lessen its belongings. This lessens the impact before the biological system's structure is on a very basic level changed to an alternate state. This isn't all around the case and there is no demonstrated connection between the species assorted variety of a biological system and its capacity to give goods and services on a manageable level.

The term ecosystem can likewise relate to human-made conditions, for example, human environments and human-affected environments, and can depict any circumstance where there is a connection between living life forms and their condition. Fewer areas on the earth today exist free from human contact, albeit some real wild regions keep on existing with no types of human intercession.

Wild is commonly characterized as a regular habitat on Earth that has not been altogether altered by human movement. The WILD Foundation really expounds, characterizing wild as: " The most intact, undisturbed wild natural areas left on our planet - those last truly wild places that humans do not control and have not developed with roads, pipelines or other industrial infrastructure." Wilderness regions and ensured parks are viewed as significant for the survival of specific species, environmental investigations, preservation, isolation, and entertainment. Some nature authors accept wild territories are crucial for the human soul and inventiveness.

"Wilderness", gets from the thought of ferocity; at the end of the day that which isn't controllable by people. The word's historical background is from the Old English wildeornes, which thus gets from wildeor meaning (wild + deor = monster, deer).

Natural life incorporates all wild plants, creatures, and different life forms. Training wild plant and creature species for human advantage has happened ordinarily everywhere throughout the planet, and majorly affects nature, both positive and negative. Natural life can be found in all biological systems. Deserts, woodlands, fields, and different regions—including the urban locales—all have particular types of wildlife.

Natural resource management refers to the management of natural resources such as land, water, soil, plants and animals, with a particular focus on how management affects the quality of life for both present and future generations (stewardship).

Environmental management is "a purposeful activity with the goal to maintain and improve the state of an environmental resource affected by human activities.

Environmental Management can be defined as "the management of the interaction and impact of human activities on the natural environment".

Natural administration further means to guarantee that biological system administrations and biodiversity are ensured and kept up for fair use by future human ages, and furthermore, keep up environment respectability as an end in itself by contemplating moral, financial, and logical (biological) factors. Natural administration attempts to recognize the elements that have a stake in the contentions that may ascend between meeting requirements yet securing the earth.

Assessment: Inspections with pertinent Government Departments (Provincial and National) we decide the degree of the issue

Control: The issue is brought to the consideration of the transgressor/polluter and the individual/organization who is educated to stop the unlawful movement and address the issue or concerns which incorporate the restoration of the regions.

Common asset the board manages dealing with the manner by which individuals and normal scenes connect. It unites land use planning, water management, biodiversity protection, and the future supportability of ventures like horticulture, mining, the travel industry, fisheries and forestry. It perceives that individuals and their jobs depend on the wellbeing and efficiency of our scenes, and their activities as stewards of the land assume a basic job in keeping up this wellbeing and profitability.

The natural habitat receives law protection, however, is regardless subject to development, inciting extreme response as esteemed scenes might be for all time adjusted. The board of the regular habitat is crucial for flexibility and alleviation anticipating characteristic occasions, for example, flooding.

Environmental resource management is an issue of expanding worry, as reflected in its commonness in fundamental writings impacting worldwide sociopolitical systems, for example, the Brundtland Commission's Our Common Future, which featured the incorporated idea of the condition and universal improvement and the Worldwatch Institute's yearly State of the World reports.

The earth decides the idea of individuals, creatures, plants, and places around the Earth, influencing conduct, religion, culture and monetary practices. It includes the administration of all segments, both living (biotic) and non-living (abiotic), and the connections among every single living species and their living spaces. Nature additionally includes the connections of the human condition, for example, the social, social and monetary condition, with the biophysical condition. The fundamental parts of natural asset the board are moral, affordable, social, and mechanical. These underlie standards and help the decision. It covers numerous territories in science, including geology, science, sociologies, political theories, open approach, biology, material science, science, human science, brain research, and physiology.

- State property: Ownership and command over the utilization of resources is in hands of the state. People or gatherings might most likely utilize the assets, however just at the authorization of the state. National backwoods, National parks and military reservations are some US models.
- Private property: Any property claimed by a characterized individual or corporate substance. Both the advantage and obligations to the assets tumble to the owner(s). Private land is the most widely recognized model.
- Basic property: It is a private property of a gathering. The group may fluctuate in size, nature and interior structure, for example, indigenous neighbors of town. Example: community forests
- Non-property (open access): There is no unmistakable proprietor of these properties. Every potential client has an equivalent capacity to utilize it as they wish. These territories are the most abused. It is said that "Everyone's property is no one's property". A model is a lake fishery.
- Hybrid: This includes all the above. A case is local vegetation management in NSW, Australia, where enactment perceives an open enthusiasm for the safeguarding of local vegetation, yet where most local vegetation exists on private land.

7.2 WHY MANAGING NATURAL ENVIRONMENT IS NECESSARY?

Is it true that we are doing what's necessary to save nature? In the event that we don't save nature, we need to confront a great deal of challenges and issues. In the event that we take a gander at the world, you see the perplexity and the disorder that we experience or face. There is a lack of water in a large portion of the nations. There are sustenance deficiencies. Individuals don't have legitimate houses to live in. The ocean level is rising. Air contamination issue is intense. We are utilizing such a large number of vehicles. Illegal medications are being sold in all nations. Liquor abuse has turned into an issue. Individuals are under mental pressure.

We should comprehend these essential standards of nature and attempt to live with nature. We ought not to depend on a lot on structures and all kind of synthetics. By saving nature, the planet and every one of its occupants is ensured a consistent supply of clean water and outside air. We regularly conflict with nature. We depend a lot on counterfeit things. We need Air-conditioned rooms. We don't care to walk. We don't care to set up our dinners. We need everything readymade.

Trees produce the oxygen that people and different animals need to inhale and assimilate the carbon dioxide that they breathe out. Trees ingest airborne toxins, for example, carbon monoxide, sulfur dioxide and nitrogen dioxide. We don't understand the significance of nature. We attempt to decimate nature. See the man-elephant strife. Perceive what number of water bodies we have filled and developed buildings; isn't this a sad circumstance?

Clean water is fundamental for every living thing. All creatures and plants are made for the most part of water and require a consistent supply of clean water to stay solid. What number of water reservoir regions have we demolished? What number of waterways and tanks have we abused and have turned out to be polluted? Overbuilt conditions disturb and overpower nature's procedures by evacuating trees and contaminating the water and air, which prompts diminished biodiversity and generally characteristic wellbeing.

We should consider nature all the more genuinely. We should regard nature. We should live near nature. We should protect wildlife. Let us not cut trees and manufacture solid structures all over the place. Animals are not crushing nature. It is we who devastate nature and make even the animals likewise endure.

Previously, the world's living and non-living systems were supported through the earth's redistribution of oxygen, nitrogen, carbon, and different components. In any case, human resource misuse in the only remaining century has essentially stressed the biological systems we depend upon for survival.

Resources are being consumed more quickly than at any time in recent memory. These incorporate wood that can be supplanted or recharged in the long run, and non-sustainable ones like minerals and unrefined oils that are in limited supply. Assets, for example, soil, which isn't recoverable in a human life expectancy once lost, are now and then mistakenly thought to be inexhaustible, making them bound to be disregarded by assurance endeavors. While even intensely collected inexhaustible assets can't be recharged rapidly enough to fulfill need, extreme utilization of non-sustainable assets can exhaust them for all time.

The environment is important because everything that supports the survival of human beings is dependent on it. It is simply the only home of mankind, making it the most important aspect of life. Wellness of the environment directly relates to the wellness and health in human beings and every other species. In addition, the environment provides remarkable beauty.

Figure 7.1: Image Symbolizing Conservation of Nature.

Source: https://medium.com/@sashajacobcanada/sasha-jacob-5-major-reasons-for-conserving-nature-c8d54bdbef56

Our nature varies for different animals but it is also suffering the same fate: the endangerment of various species and threats to our natural resources.

There are currently laws which are the solid method for Canada to follow up on their pledge to clean air, clean water, and the protection of the natural life species and their environment. That is the reason Sasha Jacob, organizer, and CEO of Jacob Capital Management Inc. worked with WWF—Canada to guarantee that the wildlife preservation and assurance are effective to help keep up the timberland environments and biodiversity in Canadian woods.

Sasha Jacob shares the reasons why we should monitor nature:

1. For wildlife assurance

   Ensuring the wildlife likewise implies that you are protecting it for what is to come. It helps in advancing biodiversity for a sound and practical environment. We people can likewise profit by ensuring wildlife on the grounds that a great deal of prescriptions can be gotten from the chemicals from nature. The preservation of their characteristic living space improves our planet and keep nature's parity.

2. For ecological assurance

   A healthy environment is basic to a sound living. The general temperature of the world's air is rising that is the reason condition security gatherings and associations are making ventures in the assurance of the earth and diminish ozone consumption.

3. For conservation

   There are species that can't make due outside their characteristic environment without human mediation, for example, that are found in a zoo or aquarium. Other migratory species are helpless against habitat space devastation. Sasha Jacob shares that the modification of a characteristic natural surrounding can result in a domino impact that damages the whole environment.

4. Natural parks

   A natural park is home to numerous endemic species and in this way should be secured. They ensure the spots of normal magnificence and it is a land saved for local plants, animals and the spots in which they live. Natural parks house a great deal of plants so they additionally have a major influence in keeping our surroundings sound.

5. For marine environments

   Our seas are the biggest environments here and they are the Earth's biggest life supportive networks. The sea likewise gives a 6th of the animal proteins that we eat and they have the most encouraging wellspring of new medicines to battle malignant growth.

The continuous obliteration of our environments and common natural surroundings has caused an irregularity in the biodiversity. Obviously, the preservation of nature depends on people to avoid the harm caused to the world by different people.

It's easy to take for the granted the importance of environmental management. However, its economic and intrinsic value cannot be overstated. Environmental management occurs at all levels from a city park's native prairie garden to grand scale projects such as management of vast wetlands like the Everglades. State and federal agencies work hand-in-hand with non-profit agencies such as the Nature Conservancy and Defenders of Wildlife to acquire and manage resources. Restoration and preservation are key components of resource management.

The Facts

The effects of habitat loss e are expansive. The U.S. Ecological Protection Agency (EPA) reports that more than 33% of the country's compromised and endangered species live just in wetlands, with about half ward upon wetlands eventually during their life cycle. Decreases in quantities of waterfowl reliant on prairie potholes have ascended with the loss of meadows as per North Dakota Game and Fish Department.

For what reason would it be a good idea for you to think about the loss of waterfowl or different creatures? Notwithstanding natural surroundings, wetlands give protection from flood. A solitary section of land of wetlands can put away to 1.5 million gallons of water. An examination by the Wetlands Initiative expresses that restoration of the 100-year flood zone on the Upper Mississippi could have abated the property harm and harvest loss of the Great Flood of 1993. Ecological administration likewise incorporates the insurance of human wellbeing through ventures, for example, the Department of Energy's cleanup of atomic waste destinations.

Proper environmental management is essential. The repercussions of blunder or no administration are extreme. With damage to habitat loss comes the danger of loss of the advantages the regular habitat gives, for example, flood protection. As non-local plants species spread into local biological systems, there are extra dangers of loss of food sources and natural surroundings.

The widespread belief is that the land will deal with itself. Without a doubt, the land and its natural life will adjust or die. At the point when changes happen too rapidly, natural life species may not recoup. On the off chance that intrusive species are permitted to assume control over wetlands, timberlands and prairies, the whole fundamental structure of the biological system is in risk. The monetary estimation of wetlands, for instance, as far as flood assurance, makes ecological administration of the most noteworthy need.

Additional benefits of environmental management include the aesthetic value of being outdoors and being able to experience the sound of flocks of geese flying or hearing river waters. It is a value that cannot be qualified but surely exists.

As Thoreau once said, "In wildness is the preservation of the world." Respect and support for environmental management practices is key to that goal.

There is much to do when it comes to rebuilding and protecting what's left of natural resources and the biodiversity within our ecosystems. Ecological preservation is an umbrella term that characterizes anything we do to secure our planet and ration its regular assets with the goal that each living thing can have an improved quality of life.

Protection works in two different ways. It is intended to secure nature by ensuring crucial resources, and it is additionally a method for living that neutralizes the reckless practices of organizations and businesses. The more cash we keep out of the pockets of enormous organizations that couldn't care less for nature, the more we can accomplish for the planet.

Environmental conservation comes in numerous ways and helps us to be careful to remember day by day decisions. Regardless of how bustling your life might be, it remains genuinely simple to make little, yet vital, changes for the benefit of the Earth. We can take little strides to advance towards real advancement. This can be accomplished by giving more consideration to what we purchase and not utilizing resources pointlessly. We should likewise make sure to reuse and discard chemicals appropriately with the goal that the ground and water systems are not harmed. The earth is crucial to our prosperity and each component ought to be ensured.

Think about all aspects of the earth, including the trees, the seas and the dirt. The trees help us to remove greenhouse gases that would make the Earth warmth up more rapidly. They hold soil set up and normally channel water that has been ingested into the ground.

The waters of the sea are unquestionably captivating, however, there is substantially more to it than visual quality. When we gaze toward the mists and feel the downpour, the sea is to thank for it. Home to innumerable sorts of marine life and plants, the sea is both a sanctuary and an actual existence source. It gives us the greater part of the Earth's oxygen supply and stores a lot of carbon dioxide to help keep the Earth cool.

The soil stands as its own supporter of life by going about as a provider of sustenance and a channel of water. Soil that is of good quality produces crops that feed people and animals. Plants and flowers grow up from the beginning of the soil, regulating our climate. The greater part of nature's water isn't made for human utilization, however, soil additionally stores groundwater and channels it, making it alright for drinking.

The trees, the oceans, and the ground; these three components are important for our reality, and there are still more sections of nature that we have to protect. This is the reason there are countless protectionists, a large number of which have committed themselves to ensure natural causes.

7.3 PROTECTING THE NATURAL ENVIRONMENT

Nature is an extremely valuable piece of our legacy: plants and soils help to purge water, timberlands go about as regular carbon sinks, and all creatures have a task to carry out in the food chain. A solid habitat is essentially significant for all eco-frameworks and it is our obligation to secure it.

- Abstain from purchasing peat items you can make your very own manure for nothing and use it in your greenhouse. Consider the chemicals you use in your home – these, in the long run, discover their way into the water seepage framework. Utilize natural options at every possible opportunity.
- Purchase natural products. No artificial fertilizers or pesticides are utilized underway, keeping them from defiling areas and entering the food chain. At the point when out for a walk, abstain from stomping on plants and do whatever it takes not to irritate wildlife.
- Be responsible with your waste– our nature isn't a dumping ground. Litter can damage wildlife and it additionally discharges poisons that can develop in nature.
- Try not to block water courses.
- Take photos home, not plants.

- Participate in wildlife surveys studies to help recognize what zones need securing. Engage with green activities or network cleanup venture.
- Change demeanors towards our common habitat – tell your companions what you know!
- People have an ethical commitment to protect nature and its highlights.
- The earth ensures the biological system.
- Ensuring nature secures mankind.
- It is an approach to provide for your age.
- Forests give crude materials to different purchaser items.
- Forests give living spaces to many animal types.
- The manner by which an individual protects the earth mirrors their character.
- Securing the earth expands the travel industry.
- The earth is viable in producing a successful food chain.
- The trees that are a huge part of the earth helps in keeping the atmosphere cool.
- A reasonable natural system results in rain.
- Trees, a huge piece of nature, assume a basic job in the refilling of waters.
- Trees assume a noteworthy job in hindering the winds.
- Trees help in keeping earth in its place.
- Numerous organisms need the earth for survival.
- Safeguarding the earth could prevent soil disintegration.
- Keeping up the earth hinders dangerous global warming.
- Keeping up the earth prompts the making of the common watersheds.
- The earth gives a fantastic wellspring of groundwater.
- The woods, a valuable piece of nature, encourage us to relax.
- The woods give us clean air.
- Protecting the earth prompts less sun based radiation.
- Protecting the earth, for example, developing of backwoods balances out noise pollution.

- Protecting the earth is basic in giving nourishment to individuals.
- Protecting the earth prompts accessibility of sustenance for the animals and plants.
- Protecting the earth advances monetary stability.
- Securing the earth satisfies us.
- Protecting the earth energizes better physical wellbeing.
- Protecting the earth reminds us to welcome the excellence of nature.
- Protecting the earth stirs our creative mind.

We just have one planet and we should all do our part to guarantee that it's secured by joining our cutting edge methods for living with more profound regard and comprehension of what nature gives. Ecological preservationists comprehend that the manner in which we live is an impression of how we feel about the normal world, and our ordinary propensities show the amount we genuinely esteem every one of the things that the Earth gives us.

Approaches

Voluntary environmental agreements

In industrial countries, voluntary environmental agreements frequently give a stage to organizations to be perceived for moving past the base administrative norms and along these lines bolster the advancement of best ecological practices. For example, in India, Environment Improvement Trust (EIT) has been working for protecting the environment and forests since 1998.

A gathering of Green Volunteers gets an objective of Green India Clean India idea. CA Gajendra Kumar Jain a Chartered Accountant is the organizer of Environment Improvement Trust in Sojat city, a little town of State of Rajasthan in India. In creating nations, for example, Latin America, these understandings are all the more ordinarily used to cure huge degrees of non-compliance with required guidelines. Because of the troubles in assessing effectiveness, their utilization is regularly questioned and, to be sure, the entire condition likely could be unfavorably influenced, therefore.

Ecosystems approach

This means to consider the complex interrelationships of a whole environment in decision making instead of essentially reacting to issues and difficulties.

In a perfect world, the basic leadership forms under such a methodology would be a cooperative way to deal with arranging and basic leadership that includes a wide scope of partners over all important legislative divisions, just as delegates of industry, natural gatherings and network. This methodology preferably underpins exchange of information, the advancement of resolution strategies and improved preservation. Religions likewise assume a significant job in the preservation of the earth.

International environmental agreements

Many of the earth's resources are especially vulnerable because they are influenced by human impacts across many countries. Thus, numerous endeavors are made by nations to create understandings that are marked by various governments to anticipate harm or deal with the effects of human action on natural resources. This can incorporate understandings that affect factors, for example, atmosphere, seas, streams and air pollution. These worldwide natural understandings are here and t legally binding documents that have lawful ramifications when they are not followed and, at different occasions, are more understandings on a basic level or are for use as sets of accepted rules. These understandings have a long history with some international agreements being set up from as right on time as 1910 in Europe, America and Africa. The absolute most understood international agreements incorporate the Kyoto Protocol.

Principle article: Kyoto Protocol

Discussion concerning environmental protection frequently centers on the job of government, enactment, and law requirement. Notwithstanding, in its broadest sense, environmental protection might be believed to be the duty of the considerable number of individuals and not just that of government. Choices that affect the earth will in a perfect world include an expansive scope of partners including industry, indigenous gatherings, environmental gathering and representatives.

Numerous constitutions recognize the key ideal to natural security and numerous global arrangements recognize the privilege to live in a sound situation. Likewise, numerous nations have associations and offices committed to ecological security. There are worldwide ecological insurance associations, for example, the United Nations Environment Program.

Albeit natural assurance isn't just the obligation of government security acts, the vast majority see these organizations as being of prime significance in building up and keeping up fundamental measures that ensure both the earth and the general population interfacing with it.

Tanzania

Tanzania is perceived as having the absolute most noteworthy biodiversity of any African nation. Practically 40% of the land has been built up into a system of secured zones, including a few national parks. The worries for the common habitat incorporate harm to environments and loss of habitat coming about because of populace development, an extension of farms or horticulture, pollution, timber extraction and huge utilization of timber as fuel.

Environmental protection in Tanzania started during the German control of East Africa (1884-1919) — colonial conservation laws for the insurance of game and woodlands were instituted, whereby confinements were put upon customary indigenous exercises, for example, hunting, wood gathering and cattle grazing. In the year 1948, Serengeti was authoritatively settled as the main national park for wild cats in East Africa. Since 1983, there has been an increasingly wide achieving exertion to oversee natural issues at a national level, through the foundation of the National Environment Management Council (NEMC) and the improvement of an environmental act. In 1998 Environment Improvement Trust (EIT) begin working for condition and forest protection in India from a little city Sojat. The originator of Environment Improvement Trust is CA Gajendra Kumar Jain working with volunteers.

The National Environment Management Council (NEMC) is an organization that was started when the National Environment Management Act was first presented in the year 1983. This chamber has the job to prompt governments and the global network on a scope of natural issues. The NEMC the accompanying purposes: give specialized counsel; arrange specialized exercises; create implementation rules and strategies; survey, screen and assess exercises that affect the earth; advance and help environmental information and correspondence; and look for progression of scientific knowledge.

The National Environment Policy of 1997 goes about as a structure for ecological basic leadership in Tanzania. The arrangement destinations are to accomplish the following:

- Guarantee feasible and fair utilization of resources without degrading the earth's wellbeing or security.
- Control the degradation of water, air, land, vegetation,
- Monitor and improve common and man-made legacy, including biological diversity of biological systems
- Improve the condition and efficiency of degraded regions
- Bring issues to light and comprehension of the connection between environment and development
- Allow for worldwide collaboration
- Use ecofriendly resources.

Tanzania is a signatory to a noteworthy number of worldwide associations including the Rio Declaration on Development and Environment 1992 and the Convention on Biological Diversity 1996. The Environmental Management Act, 2004, is the primary complete lawful and institutional structure to direct environmental-management decisions.

China

Following the 1972 United Nations Conference on the Human Environment held in Stockholm, China started setting up environmental protection agencies and putting controls on a portion factory waste. China was one of the main creating nations to execute a sustainable development strategy. In 1983 the State Council declared that environmental protection would be one of China's essential national approaches and in 1984 the National Environmental Protection Agency (NEPA) was set up. Following extreme flooding of the Yangtze River bowl in 1998, NEPA was moved up to the State Environmental Protection Agency (SEPA). In 2008, SEPA wound up known by its present name of Ministry of Environmental Protection of the People's Republic of China (MEP).

China has been one of the top performing nations as far as GDP development (9.64% in the previous ten years). Nonetheless, the high monetary development has put tremendous weight on its condition and the environmental difficulties that China faces are more noteworthy than general nations. In 2010 China was positioned 121st out of 163 nations on the Environmental Performance Index.

Quick development in GDP has been China's fundamental objective during the previous three decades with mainly wasteful resource use and high pollution to accomplish high GDP. For China to grow reasonably, environmental protection ought to be treated as an indispensable piece of its financial policies.

China's investment in renewable energy reached $15.6 billion or 18% in 2007 making it ~10% of the global investment in this area while expenditure for the environment was at 3.4 times from 2000.

Quote from Shengxian Zhou, head of MEP (2009): "Good economic policy is good environmental policy and the nature of environmental problem is the economic structure, production form and develop model."

European Union

Environmental protection has turned into a significant errand for the organizations of the European Community after the Maastricht Treaty for the European Union endorsement by all the Member States. The EU is as of now exceptionally dynamic in the field of ecological strategy with directives like those on environmental information.

Russia

In Russia, environmental protection is viewed as an indispensable portion of national wellbeing. There is an approved state body - the Federal Ministry of Natural Resources and Ecology. Be that as it may, there is a great deal of environmental issues in Russia.

Latin America

The United Nations Environment Program (UNEP) has recognized 17 megadiverse nations. This incorporates six Latin American nations: Brazil, Colombia, Ecuador, Mexico, Peru and Venezuela. Mexico and Brazil emerge among the rest since they have the biggest areas and biodiversity. These nations are a noteworthy worry for environmental protection since they have high rates of deforestation, habitat loss, pollution, and population.

Brazil

Brazil has the biggest extent of the world's tropical forests, 4,105,401 km^2 (48.1% of Brazil), amassed in the Amazon locale. Brazil is home to tremendous natural decent variety, first among the megadiverse nations of the world, having between 15%-20% of the 1.5 million all inclusive species

The association responsible for protection is the Brazilian Ministry of the Environment (in Portuguese: Ministério do Meio Ambiente, MMA). It was first made in the year 1973 with the name Special Secretariat for the Environment (Secretaria Especial de Meio Ambiente), changing names a few times, and receiving the last name in the year 1999. The Ministry is in charge of tending to the accompanying issues:

- A national approach for the earth and for water assets; An arrangement for the protection, preservation and economical utilization of biological systems, biodiversity and forests;
- Proposing techniques, components, financial and social tools for improving quality, and manageable utilization of natural resources;
- Approaches for incorporating production and the earth;
- Natural approaches and projects for the Legal Amazon; Ecological and financial regional zoning.

In 2011, protected areas of the Amazon secured 2,197,485 km^2 (a territory bigger than Greenland), with protection units, similar to national parks, representing simply over half (50.6%) and indigenous domains speaking to the staying 49.4%.

Mexico

With more than 200,000 distinct species, Mexico is home to 10–12% of the world's biodiversity, first in reptile biodiversity and second in mammals — one report shows that over half of all animal and plant species live in Mexico.

The historical backdrop of the ecological approach in Mexico began during the 1940s with the institution of the Law of Conservation of Soil and Water (in Spanish: Ley de Conservación de Suelo y Agua). After three decades, toward the start of the 1970s, the Law to Prevent and Control Environmental Pollution was made (Ley para Prevenir y Controlar la Contaminación Ambiental).

In the year 1972 was the main direct reaction from the national government to address health effect impacts from environmental issues. It set up the authoritative association of the Secretariat for the Improvement of the Environment (Subsecretaría para el Mejoramiento del Ambiente) in the Department of Health and Welfare. The Secretariat of Environment and Natural Resources (Secretaría del Medio Ambiente y Recursos Naturales,

SEMARNAT is Mexico's condition service. The Ministry is in charge of tending to the accompanying issues:

- Advance the assurance, reclamation, and protection of biological systems, resources, and environmental administrations and encourage their utilization and reasonable improvement.
- Create and execute a national strategy on natural resources
- Advance environmental management inside the national region, in a joint effort with all degrees of government and the private sectors.
- Assess and give assurance to the natural effect explanations for environmental impact statements and the anticipation of environmental harm
- Look at the security of the ozone layer.
- Direct work on national meteorological, climatological, hydrological, and geohydrological frameworks, and take an interest in worldwide conventions regarding these matters.
- Manage and screen the protection of water systems

Oceania

Australia

In 2008, there was 98,487,116 ha of the terrestrial protected area, covering 12.8% of the land area of Australia. The 2002 figures of 10.1% of the terrestrial area and 64,615,554 ha of the protected marine area were found to poorly represent about half of Australia's 85 bioregions.

Natural assurance in Australia could be viewed as beginning with the development of the first National Park, Royal National Park, in 1879. Increasingly dynamic natural insurance had it begin during the 1970s with significant worldwide projects, for example, the United Nations Conference on the Human Environment in 1972, the Environment Committee of the OECD in 1970, and the United Nations Environment Program of 1972. These occasions established the frameworks by expanding open mindfulness and backing for guideline. State ecological enactment was sporadic and inadequate until the Australian Environment Council (AEC) and Council of Nature Conservation Ministers (CONCOM) were set up in 1972 and 1974. These boards have since been supplanted by the Australian and New Zealand Environment and Conservation Council (ANZECC) in 1991 lastly the Environment Protection and Heritage Council (EPHC) in 2001. The Environment Protection and Biodiversity Conservation Act 1999 is the main protection legislation in Australia.

The act covers eight main areas:

- National Heritage Sites
- World Heritage Sites
- RAMSAR wetlands
- Nationally endangered or threatened species and ecological communities
- Nuclear activities and actions
- Great Barrier Reef Marine Park
- Migratory species
- Commonwealth Marine areas

Examples are: Kakadu National Park (partnerships with traditional native owners) and Australian Alps National Parks and Reserves (co-operative management).

At a state level, the majority of security issues are left to the obligation of the state or region. Each state in Australia has its own environmental protection enactment and organizations. Most grounds are overseen by states and domains with state administrative acts making various degrees and meanings of protected regions, for example, wild, national land and marine parks, state backwoods, and preservation regions. States additionally make a guideline to constrain and regulate air, water, and sound pollution.

Australia positions second spot on the UN 2010 Human Development Index and one of the most reduced debt to GDP ratios of the created economies. This could be viewed as coming at the expense of nature, with Australia being the world chief in coal exportation and extinctions.

New Zealand

At a national level, the Ministry for the Environment is in charge of natural strategy and the Department of Conservation tends to preservation issues. At a provincial level, the territorial boards oversee the enactment and address local environmental issues.

Switzerland

The natural security in Switzerland is for the most part dependent on the measures to be taken against Earth-wide warming. The pollution in Switzerland is, for the most part, brought about by vehicles and the literation by visitors

United States

Since 1969, the United States Environmental Protection Agency (EPA) has been working to protect the environment and human health. Each U.S. state has individual state departments of environmental protection.

The EPA has drafted "Seven Priorities for EPA's Future", which are:

"Taking Action on Climate Change"

"Improving Air Quality"

"Assuring the Safety of Chemicals"

"Cleaning Up Our Communities"

"Protecting America's Waters"

"Expanding the Conversation on Environmentalism and Working for Environmental Justice"

"Building Strong State and Tribal Partnerships"

7.4 WAYS TO MANAGE THE ENVIRONMENT

Integrated natural resource management (INRM) is a process of managing natural resources in a systematic way, which includes multiple aspects of natural resource use (biophysical, socio-political, and economic) meet production goals of producers and other direct users (e.g., food security, profitability, risk aversion) as well as goals of the wider community (e.g., poverty alleviation, welfare of future generations, environmental conservation). It focuses on sustainability and at the same time tries to incorporate all possible stakeholders from the planning level itself, reducing possible future conflicts. INRM is being utilized widely and been effective in both local and community-based common administration. As far as understanding, Young underscores some significant purposes of land management:

- Understanding the procedures of nature including biological system, water, soils

- Utilizing suitable and adjusting the management systems
- Participation between researchers and local dwellers who have information and abilities

Dale et al. (2000) described five key and accommodating natural standards for the land manager and people. The environmental standards identify with time, place, species, and disturbances.

- Inspect the effects of local decisions, and the impacts on resources.
- Plan for long term change.
- Save rare landscape and related species.
- Stay away from practices that exhaust natural resources.
- Hold huge land that contains critical habitats.
- Limit the spread of non-native species.
- Keep away from or make up for the impacts of advancement on the environment.
- Execute land-use and land- management practices that are in sync with the zone.

Objectives to serve nature, communicated by researchers and environmentalists include:

- End contamination and toxicants in air, water, soil, structures, , and food.
- Conservation of biodiversity and protecting threatened species.
- Protection and reasonable utilization of resources, for example, water, land, air, etc. .
- Ending human-induced warming, that is a danger to biodiversity, and a risk to human populace.
- Moving from petroleum derivatives to sustainable power source in power, warming and cooling, and transportation,.
- Moving from meat-eating regimens to generally plant-based eating regimens so as to help moderate biodiversity loss and environmental changes.
- Administration for waste decrease (or even zero waste), reuse, reusing, fertilizing the soil, composting, and anaerobic assimilation of sewage muck.
- Moderating and adjustment of human development.
- Ration by alternative energy sources.

- Sunlight based vitality and wind power are two of the sustainable power source alternatives that we could utilize all the more regularly. While we don't totally control how energy is offered to us, there's dependably space to live such that utilization less of the energy that can't be renewed. Start to ration by rolling out little improvements to regular schedules. For instance, use window light as opposed to turning on lights during the day and supplant standard lights with bulbs that are energy effective.
- Deforestation keeps on being a noteworthy ecological issue. This affects the ozone and loss of biodiversity and habitats.
- Planting trees is an approach to give back in light of the fact that it helps in the rebuilding of homes for wildlife, food sources, and restorative properties that the trees give. As trees develop, they shield the soil from climate conditions and shield us from excess carbon dioxide, empowering us to live more and all the more easily.
- Help to ensure soil quality via compost. This increases plant abundance and its associated benefits.

Figure 7.2: Deforestation.

Source: https://www.conserve-energy-future.com/methods-and-importance-of-environmental-conservation.php

Practice habits that are part of conservational efforts that are given below:

- Keep your vehicle fit to curtail the carbon that it discharges into the climate. When buying vehicles look past the external appearance and think about the ecological effect. Choose to buy electric cars. Walk, ride a bicycle, or utilize shared transportation when conceivable.
- Begin growing for your wellbeing purposes. This avoids the contamination of using pesticides or toxic chemicals.
- Converse with others about environment protection. There are numerous fun approaches to do this. For instance, you can host parties dependent on eco-topics to begin significant discussions. You can utilize occasions such as Earth Day to acquaint others with reasonable practices that can profit them just like the planet. The more individuals that perceive the ecological issues we face, the more probable they are to participate in positive improvements.

Ecological Conservation: 8 Additional Ways You Can Help

- Quit smoking. On the off chance that you don't smoke, you decrease air contamination and improve your wellbeing.
- Change the manner in which you clean the house. Utilize reasonably made things or every regular item that isn't made with risky chemicals.
- Ration water. Wash your garments and dishes simply after a full burden.
- Unplug things. On the off chance that a thing isn't being used, that doesn't mean it isn't requiring power. At the point when a thing is unplugged, it's not utilizing any power whatsoever. This spares more vitality for some time later and saves costs.
- Figure out how to upcycle. Research to learn fun new tasks that can carry new life to old things with the goal that you don't need to discard them.
- In the event that you can, abstain from utilizing plastic water jugs and paper so frequently. Anything you do utilize that is recyclable, make certain to discard it appropriately.
- Contact government authorities and support petitions and different organizations that advance change.

- Be benevolent to other people. Give things that you don't utilize any longer as opposed to discarding them. You abstain from destroying profitable things while making it accessible to another person for less cash. On a comparative note, buy things that were previously owned as opposed to purchasing new material that has been mass created with no respect for nature.

REFERENCES

1. https://en.wikipedia.org/wiki/Natural_resource_management
2. https://www.rtpi.org.uk/events/events-calendar/2017/april/managing-the-natural-environment/
3. https://france.arocha.org/en/our-work/managing-natural-environments/
4. https://www.eac.int/environment
5. https://en.wikipedia.org/wiki/Environmental_resource_management
6. https://en.wikipedia.org/wiki/Natural_environment
7. http://westcoastdm.co.za/about/environmental-management/
8. https://online.unity.edu/understanding-environment-context-sustainability-science-natural-resource-management-masters/
9. http://www.ft.lk/other-sectors/it-is-important-to-preserve-nature/57-572109
10. https://medium.com/@sashajacobcanada/sasha-jacob-5-major-reasons-for-conserving-nature-c8d54bdbef56
11. https://en.wikipedia.org/wiki/Environmental_issue
12. https://en.wikipedia.org/wiki/Environmental_protection
13. https://www.conserve-energy-future.com/methods-and-importance-of-environmental-conservation.php
14. http://www.askaboutireland.ie/enfo/sustainable-living/how-to/protect-our-natural-envir/
15. https://www.quora.com/Why-is-preservation-and-conservation-of-the-environment-important
16. https://www.slideshare.net/lalindaperera/why-natural-environment-is-important-in-urban-planning
17. https://www.hunker.com/12216593/the-importance-of-environmental-management
18. https://www.conserve-energy-future.com/15-current-environmental-problems.php
19. https://www.dw.com/en/five-of-the-worlds-biggest-environmental-problems/a-35915705
20. https://www.renewableresourcescoalition.org/top-environmental-problems/

Environmental Issues

8

"Address these environmental issues and you will address every issue of man. And we keep dabbling in things that aren't really that important in the long term."

Ted Danson

CONTENTS

We are seeing and hearing a lot about the Environment. It is the environment that gives us all. But today, there are many environmental issues that affect society and the world as a whole. So much attention is being given to the environment these days because men have started realizing their mistakes after seeing the impact of their inappropriate behavior towards the environment. This Chapter is dedicated to deeply analyze the various environmental issues that the world is facing currently and how men have contributed to these issues. We also look at the various reasons for this environmental crisis.

8.1 DIFFERENT ENVIRONMENTAL ISSUES

Our Planet Mother Earth is currently facing a lot of environmental issues. Environmental issues are the result of the harmful effects of human activity on the biophysical environment. The past few decades have seen an increased exploitation and degradation of our planet. Some of the environmental problems like air pollution, global warming, acid rain, ozone layer depletion, urban sprawl, waste disposal, water pollution, climate change cause an effect on every living organism and the planet. Turning a blind eye to the signals from nature has resulted in many calamities.

Figure 8.1: Destruction of Environment.

Source: https://www.asente.ch/environmental-issues/

He can see Climate disasters more often. Around 70 % of the natural disasters are now climate-related and these have increased from around 50

% when compared two decades ago that has taken a huge toll. In the last decade alone, 2.4 billion people were affected by climate related disasters, compared to 1.7 billion in the previous decade. Along with taking human life, these disasters create a lot of tension in the economy. The disaster management cost has risen tenfold between 1992 and 2008. The forecasts suggest an increase in such disasters in the forthcoming years. Intense tropical storms, heavy rains, repeated flooding and droughts are projected to increase in the absence of appropriate action.

Humans are the main cause for environment destruction and this is a global problem that is occurring every day. According to Living Blue Planet 24, by the year 2050, the global human population is expected to grow by 2 billion people, thereby reaching a level of 9.6 billion people and so the activities of humans are only going to increase. According to the report, "Our Changing Climate",

"The global warming that has been going on for the past 50 years is primarily due to human activities. Temperature comparisons have given us some solid proof of how the temperature has drastically increased over the past years. The U.S. average temperature has increased from 1.3 °F to 1.9 °F since 1895, with most of the increase taken place since around year 1970".

Current environmental issues also include climate change, environmental degradation, pollution, resource depletion etc.,

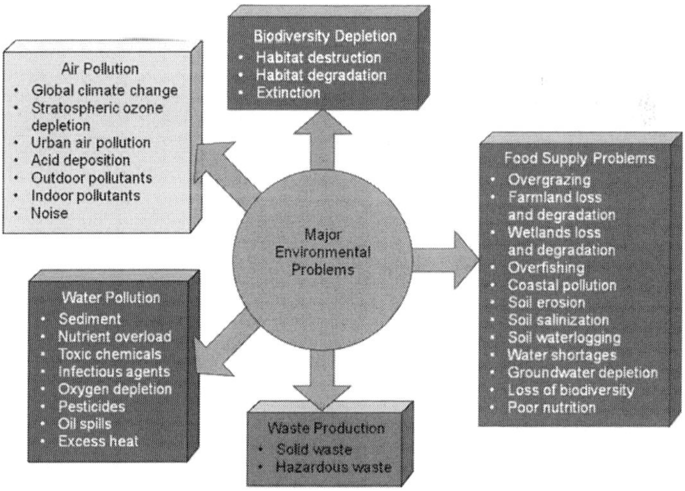

Figure 8.2: Major Environmental Problems.

Source: http://environmentalscience107.blogspot.com/2014/02/basic-environmental-issues.html

Let us now see in detail the different environmental concerns that the world is facing today:

Pollution

Pollution is a great problem faced by the world today. Pollution comes in many forms. Air, soil, water and land all have the capability to be polluted. Pollution is hazardous both to the people and the environment. Contaminated waters become undrinkable. Polluted air weakens the ozone layer and causes health problems and also gives breathing difficulties. Contaminated soil destroys habitats, irrigation and results in less crop cultivation.

Figure 8.3: Effects of Pollution.

Source: https://www.indiacelebrating.com/environmental-issues/effects-of-soil-pollution/

Pollution is said to be one of the primary causes of most of the other environmental problems, including climate change and biodiversity. There are seven major types of pollution affecting our environment i.e., air, water, soil, noise, radioactive, light and thermal and all these types of pollution. All these types of pollution and environmental problems are interlinked with one another and they influence one another. So, to prevent one type of pollution we have to work on tackling all of them.

The restoration of these systems can entail many millions of years. There are many factors aiding this. The gases coming out of industry and motor vehicles are the number one pollutants. Heavy metals, nitrates and plastic are the toxins that are responsible for pollution. Oil spills and runoffs contribute to water pollution while emitted gases make up air pollution and waste deposited constitutes soil pollution.

Let us get deeper into the different types of pollution:

Air Pollution

The air present in the atmosphere is contaminated. Many unwanted and even poisonous elements are getting mixed up in the atmosphere which causes air pollution. The air in the atmosphere is overloaded with Carbon dioxide. The concentration of Carbon dioxide in the atmosphere has raised from 280 parts per million (ppm) 200 years ago, to about 400 ppm today. This is a huge increase and it is outrageous. This is caused by deforestation for agriculture, burning of fossil fuels , and industrial activities that release poisons.

Air pollution is becoming a dangerous problem, especially in metro cities. As per the data released by The World Health Organization (WHO), 80% of people living in urban areas are exposed to polluted air considered unfit. There are also many other elements constituting to air pollution. The increasing amount of Carbon dioxide from burning fuels is one of the main reasons for air pollution. The World Health Organization estimated that one in nine deaths in 2012 were attributable to diseases caused by carcinogens and other poisons in polluted air.

Figure 8.4: Air Pollution Harm Humans.

Source: https://nation.com.pk/29-Aug-2018/air-pollution-may-harm-cognitive-intelligence

The human body is majorly affected by air pollution because it is what we breathe. Air pollution even affects animals and the environment. The garbage that we dump on a daily basis is also a cause of air pollution because this garbage releases toxic gases when they are burnt. The fumes coming out of industries and vehicle engine vehicles also pollute the air and these are the most obvious toxins.

Air pollution also accounts for other environmental issues, like acid rain and eutrophication. We will discuss them in detail in the below sections. Whatever happens in the environment has several other effects because they are all linked to each other.

Water Pollution

Water pollution is not only a huge concern for human beings but it also takes a toll on our environment. Polluted water kills everything because everything in the universe needs water for survival. This involves oil spills, draining of plastic waste and toxic chemicals our waterways. Water sources also get polluted due to acid rain and urban sprawl; we will see more about them in the later sections.

There are many sources of fresh water available in the universe because it is crucial to life on Earth. But increasing human activities pollute these freshwater sources. The main thing that gets affected badly due to water pollution is the marine life. Where will the fishes and other marine creatures go if their only habitat is polluted? Many of the marine creatures are becoming extinct and sometimes they face altering reproductive cycles.

Figure 8.5: Sources and Effects of Water Pollution.

Source: https://www.kent.co.in/blog/water-pollution-and-its-smart-purification-process/

The presence of Carbon at a higher rate in the atmosphere and water is the cause of major problems. Carbon in water targets shellfish and microscopic fish akin to osteoporosis in humans.

Getting clean drinking water is tedious as it is turning into a rare commodity making it a monetary and political concern since humans' start fighting for their basic need. Industrial and agricultural wastes target water which affects all forms of life as mentioned.

The world is focusing more on the carbon emission and the effects of the carbon cycle that they fail to notice the effects of human use of nitrogen.. When the nitrogen level in water increases over stimulates plant and algae growth that lowers air and light entering the water.

There is also water scarcity in many places and this issue is just becoming serious. Saltwater, freshwater, groundwater and surface water are the basic wellsprings of water. There are two different ways water can be scarce: the absence of water and the absence of drinkable water.

Water scarcity often happens since we do not give adequate respect for water since its importance is not felt. We take all the natural resources for granted. People often leave the tap on after taking water from it or does not close it properly. We can see on the streets water leaking out of taps and no one cares to go and turn off the handle. This is the main reason why many cities are running out of water.

People are spending money to buy good and clean drinking water. The contamination of potable water with chemicals or toxins or microbes is a serious issue. According to data facts, approximately 780 million people have no access to clean water at all. We should not assume that these happen only in under-developed countries or developing countries. An example is the 2017 drought in California

Soil and Land Pollution

Land and pollution is the degradation of the earth's surface and is also an environmental issue with the capacity to make a huge impact by turning into air pollution and soil pollution. It is due to human activities like industries, agriculture, littering, mining, and deforestation.

Figure 8.6: Land and Soil Pollution.

Source: https://helpsavenature.com/facts-about-land-pollution

The Soil is polluted by all the mechanical and industrial waste and garbage that takes essential supplements out of the soil.

Agricultural Pollution

The way of farming and irrigation has changed a lot. Modern day agriculture uses chemical products like pesticides and fertilizers to control and kill local pests. These pesticides and fertilizers are very harmful to the plants and the crops as well.

Figure 8.7: Agricultural Pollution.

Source: https://weed-killer.reliongroup.com/content/settlement/

Some of the chemicals when sprayed do not disappear and in-fact seeps into the ground and thereby harms plants and crops. Agricultural pollution also occurs when contaminated water is used for irrigation purpose by farmers.

Light and Noise Pollution

Noise pollution refers to temporary disruption due to large unpleasant noise. Many human activities cause noise pollution and this is only increasing day-by-day. Growth of industrialization, construction activities, increase in a number of vehicles and traffic, lack of planning transportation systems are few of the causes of noise pollution.

Figure 8.8: Vehicles Create Noise Pollution.

Source: https://www.thedrum.com/news/2017/06/20/delhi-traffic-police-disapproves-body-wrap-ads-buses

Celebrations, festivals, change of human lifestyle are all causing overuse of lights even during night times and this causes light pollution.

Climate Change

The biggest environmental concern that is alarming everyone today is the Climate change. Many countries are debating on Climate change. Climate change is not something that we are anticipating. The world has already started seeing the destructive consequences of it which include prolonged or shortened seasons, melting of polar ice, change in general climate situation, frequent floods, thunderstorms, wildfires, super hurricanes, droughts, the spread of new diseases etc.,

There are many reasons for climate change; the most important one is global warming. Global Warming occurs due to the increase in atmospheric temperature and this is caused mainly by the burning of fossil fuels and the release of harmful gases by industries. There are many studies conducted to identify the reasons for climate change and 97% of scientists agree that greenhouse gasses, both natural and those created by humans, are the main cause for global warming and climate change.

There is a rise in both the Earth's temperature and the sea levels. As per the Intergovernmental Panel on Climate Change Report, the carbon dioxide equivalent of greenhouse gases (GHG) in the atmosphere has already exceeded 400 parts per million ("long-term" GHG exceeding 455 parts per million) referred to as a tipping point.

Figure 8.9: Image Depicting the Impact of Climate Change.

Source: https://arstechnica.com/science/2018/03/here-are-answers-to-a-federal-judges-queries-about-climate-science/

The rise in ocean temperature is affecting sea life and ecosystems habituated there. Melting of ice caps is increasing the sea level and the rises in global sea levels are shrinking our land, causing mass floods and freak weather incidents across the world. The prolonged occurrence of these events will destroy the world and nothing can be done to reverse it.

The world is already at the risk of many areas of pollution. There's nothing much left. According to The UN Office for the Coordination of Humanitarian Affairs (OCHA),

Climate change is not just a distant future threat. It is the main driver behind rising humanitarian needs and we are seeing its impact. The number of people affected and the damages inflicted by extreme weather has been unprecedented. Greenhouse gasses are the main cause of climate change, specifically those emitted from the human population. This has an impact on habitats, agriculture, the ocean and natural disasters."

Global Warming

We have seen previously that global warming is the main reason for climate change. Global Warming is creating a lot of problems and we are also a part of it. All these problems are causing destruction in various ways. Global Warming is caused by the increase in the emission of greenhouse gases. Human practices have resulted in the emission of Greenhouse gases. Burning of fossil fuels causes the discharge of greenhouse gasses leading to environmental changes.

Figure 8.10: Global Warming Can Destroy Earth.

Source: https://www.geospatialworld.net/blogs/global-warming-climate-change-and-geospatial-technologies/

Global warming boosts Earth and sea temperatures to increase the sea levels, cause the melting of ice caps to lead to flash floods, excessive snow or desertification. Temperature increases, like climate change, are the consequence of human practices, including the use of greenhouse gasses. Such changes in the atmosphere thus damage our home planet.

Deforestation

Deforestation is a huge environmental problem and this will just continue to get worse. Humans are cutting down forests for urbanization. In the tropical region, species-rich wild forests are being destroyed, often to make way for cattle ranching, soybean or palm oil plantations, or other agricultural monocultures. Illegal logging of woods and cutting down of forests for timber and other high-quality woods for commercial purpose are also some reasons for deforestation. Due to an increase in temperature, wildfires occur naturally and destroy the forests.

The increase in population has also resulted in increasing levels of deforestation since people destroy forests to create their habitat. The tropics see the destruction annually of about 7.3 million hectares (18 million acres) of forest. Earlier the tropical forests used to cover about 15 % of the land but now they are down to 6 or 7 % that is subject to further damages.

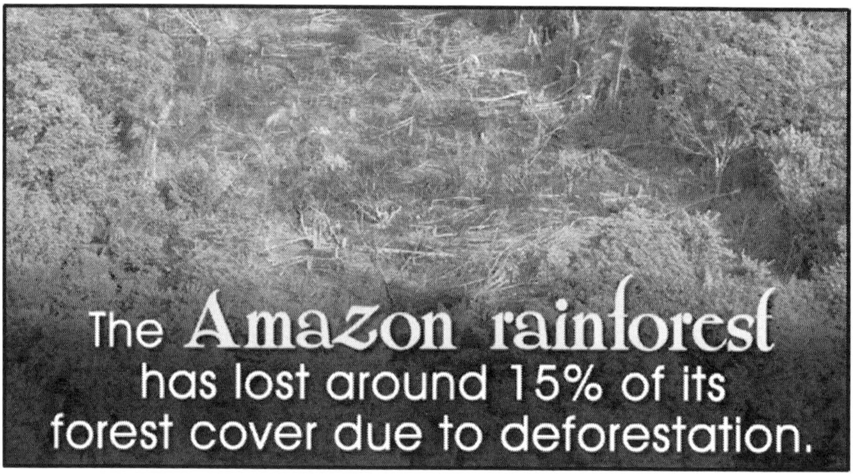

Figure 8.11: Deforestation.

Source: https://helpsavenature.com/deforestation-statistics

Forests and trees are of great benefit to the universe and essential for survival. The benefits of forests in supplying food, oxygen and carbon sinks and other necessities have been discussed earlier. Deforestation is removing the trees for construction or commercial interests.

Deforestation has many side effects which people do not understand. We all talk about the emission of greenhouse gases by burning of fossil fuels but everyone should know that at least 15 % of greenhouse gas emissions don't come from cars or factories but they come from deforestation. Many species lose their habitat due to deforestation causing the extinction of many plant and animal species to change the climate and kill ecosystems.

Genetic Modification

Environmental issues are getting more and deeper and also problematic. The technological advancements are causing more harm to the environment. Genetic modification utilizing biotechnology is called genetic engineering. Many new crops and hybrid varieties are produced by this method. Yet, the exact effects of such genetic engineering of food in terms of the environment and consumers remain controversial.

Man-Made chemicals bring more environmental issues. For example, the Monarch butterfly population in the United States has dipped by 90% reduction due to glyphosate in weed killers. Other speculations are the possible release of chemicals from genetically-modified plants to affect soil microbes thus possibly endangering the lives of future generations.

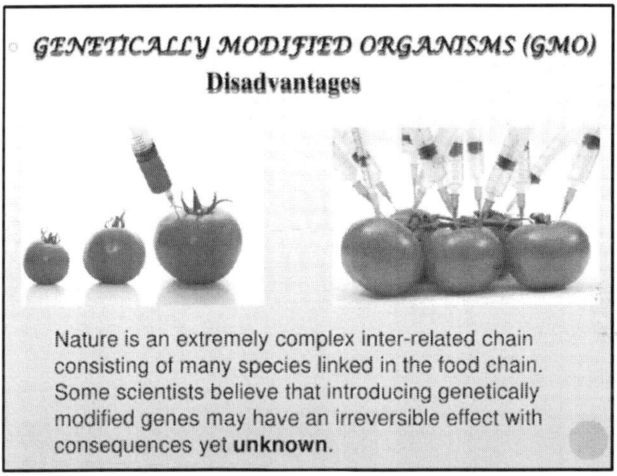

Figure 8.12: Genetically Modified Organisms (GMO).

Source: http://ww1.nora.pasalo.info/creative-writing/id-2479

There are many things that are changing in our environment. Many studies conducted demonstrate the occurrence of these changes and all these changes are having an effect on life around us. However, most of us would be unaware of the specific issues that have led to these changes. Terms like "climate change" and "genetic modification" are some of the reasons that cause changes in the environment. Most of these issues are linked to one another and that is what makes it even more difficult and challenging.

Overpopulation

Our planet Earth is becoming more and heavier as the residents of the planet are constantly increasing.

The growing population is in no way good to the environment. As the population increases, the natural resources consumption increases and so there is a deficit in the availability of natural assets like water, fuel and food. Overpopulation is becoming one of the most important environmental concerns. Without adequate natural resources, the survival rate may decrease in the future. The population will be able to cope up if there is nothing to eat or drink. People must realize this before it gets too late.

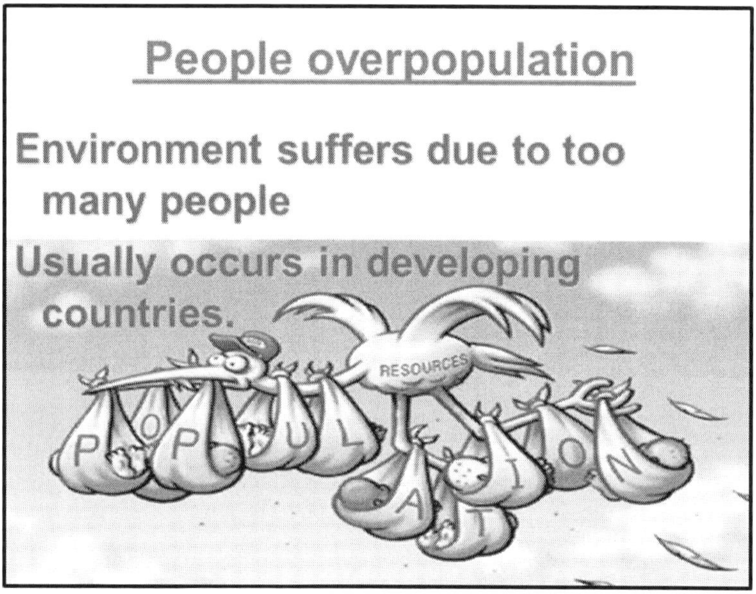

Figure 8.13: Overpopulation Effects.

Source: https://slideplayer.com/slide/6968253/

Overpopulation does not only lead to a reduction in natural resources; there are also many other consequences. More people mean release of more Carbon dioxide in the atmosphere. So these account to more greenhouse gases which lead to climate change. We already knew the negative effects of climate change.

Public Health Issues

All the environmental issues are a great threat to the human community. All other creatures are also not exempted from this. Most of the environmental problems are centered on the health of the public. According to the WHO, nearly 25% deaths each year are directly due to unhealthy environments.

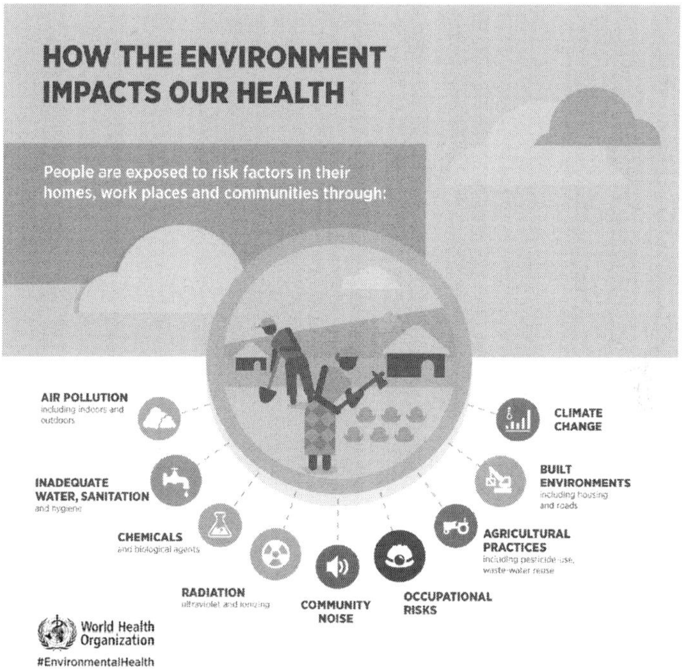

Figure 8.14: How the Environment Impacts our Health.

Source: https://www.neha.org/about-neha/definitions-environmental-health

Due to the current environmental problems, even the health of animals, birds and other creatures are at high risk. The presence of chemicals and pathogens increases the risk of diseases and climate changes that further quicken the spread of diseases.

Loss of Biodiversity

The importance of biodiversity has been specified in the initial sections of this book.

According to the World Wildlife Federation, in the last three decades alone biodiversity has declined to 27 %. There are various threats that cause a decline in biodiversity and that includes urban sprawl, deforestation, overpopulation, pollution, global warming and climate change. Destruction of habitat is a leading cause for biodiversity loss. There is a prediction of a 6th mass extinction by scientists that in turn would affect us adversely.

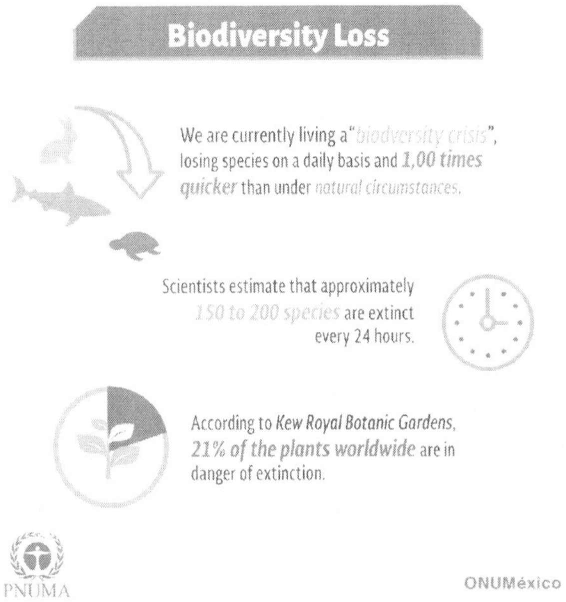

Figure 8.15: Loss of Biodiversity.

Source: http://www.teebweb.org/teeb-mexico/biodiversity/

It took billions of years for the ecosystem to evolve and now all the expanding human activities have put them in great danger. This reduced biodiversity alters the equilibrium of food chains and ecosystems hence affecting the future generations.

Many natural processes in the environment such as pollination is very crucial to the survival of the eco-system but human activity threatens the same. Another example is the destruction of coral reefs that harbor life in many oceans.

Household and Industrial Waste

Accumulation of waste or garbage is one of the most growing environmental problems arising from resource exploitation. The developed countries are producing an excessive amount of waste or garbage and dumping their waste in the oceans and, less developed countries. Most of the times, more than the dumping of waste, transfer of waste is creating more problem. Most developed countries are well-versed in dumping large quantities of waste or junk in seas and oceans. This is called littering.

Littering is the disposal of debris improperly. Disposal of waste is one of the serious problems that the world is facing today. The impacts of littering are huge. This causes damage to the ecosystems and involve money in cleaning. There are many kind of wastes and everything has its own danger.

The hazard of nuclear waste it is known. Other wastes include electronics, plastics and waste from healthcare or medical waste.

Figure 8.16: Types of Waste.

Source: https://www.slideshare.net/anshikamehrotra7/generation-of-waste-71694726

There is another type of garbage which is called the landfills. Landfills are huge garbage dumps that produce toxic gases and make a place look ugly. The trash thrown out by humans ends up either in landfills or in their habitats or oceans. Garbage dumped in oceans pose a great danger to marine life. The marine species mistake garbage for food and get tangled in it.

Burning waste only help in increasing the problem as toxins are emitted. Methane (a greenhouse gas) is generated in a landfill that is associated with explosion potential.

Ozone Layer Depletion

The Ozone layer is very important for our planet. It is an undetectable layer of protection around the planet that secures us from the sun's unsafe beams. The Ultra Violet rays emanating from the Sun is very harmful and it is the Ozone layer that prevents this harmful layer from reaching the earth. The Ultra Violet rays can cause eye diseases and skin diseases such as skin cancer. It even affects plant life; the UV rays is said to reduce plankton in marine environments. This Ozone layer is now on the verge of danger as many environmental problems are causing it to deplete.

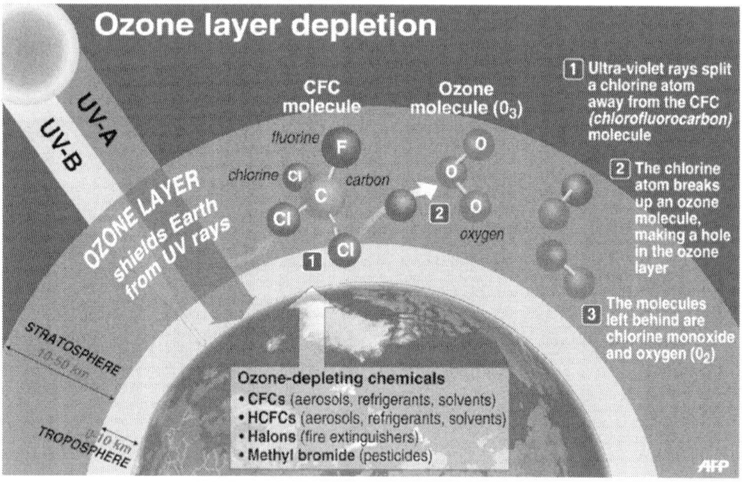

Figure 8.17: Ozone Layer Depletion.

Source: http://www.eaglenews.ph/ozone-layer-declining-over-populated-zones-study/

Depletion of the critical Ozone layer of the air is caused due to the contamination brought about by Bromide and Chlorine found in Chlorofloro Carbons (CFC's). When these poisonous gases reach the upper parts of the atmosphere, they cause a hole in the ozone layer. The biggest gap in the Ozone layer is said to be over the Antarctic region. The CFC's are banned in many industries and consumer products. This is one of the most important current environmental problems. A single atom of CFC can destroy thousands of ozone molecules.

Mining

Mining is the process of extraction of minerals from Earth's core. Many toxic gases are released during the process of mining and these toxic emissions can cause air, water and soil pollution. Sometimes these toxic gases even take human lives.

Figure 8.18: Mining.

Source: https://thinksask.ca/invest/mining

Natural Resource Depletion

Natural resource depletion is another crucial current environmental problem. The energy resources available in the environment are constantly being used for ages now. Humanity has no control over using these resources as they may get over while the burning of fossil fuels creates previously mentioned issues ..

Figure 8.19: Image Depicting Depletion of Natural Resources.

Source: https://www.123rf.com/stock-photo/resource_depletion.html?sti=lh5yr2l12uk6o6qwfv|

Increased resource use is linked to a number of other environmental issues, such as air pollution and population growth. Over time, the depletion of these resources will lead to an energy crisis, plus the chemicals emitted by many natural resources are strong contributors to climate change. Globally, people are taking various efforts to transform the usage to renewable sources of energy like solar, wind, biogas and geothermal energy.

Nuclear Issues

Figure 8.20: Nuclear Radiation.

Source: https://www.aether-force.com/the-radiation-hoax-nuclear-scientist-eats-drinks-and-swims-in-radiation/

Improper disposal of radioactive waste can cause toxic effects on life all around.

Loss of Endangered Species

Our planet Earth will potentially run out of species if the environmental problems are not resolved. Due to the list of environmental issues happening on this planet, both ecosystems and species are affected. For example, the rising temperatures of the Arctic cause the sea ice to melt and this eliminates the habitat of polar bears. The list of endangered species includes other animals as well.

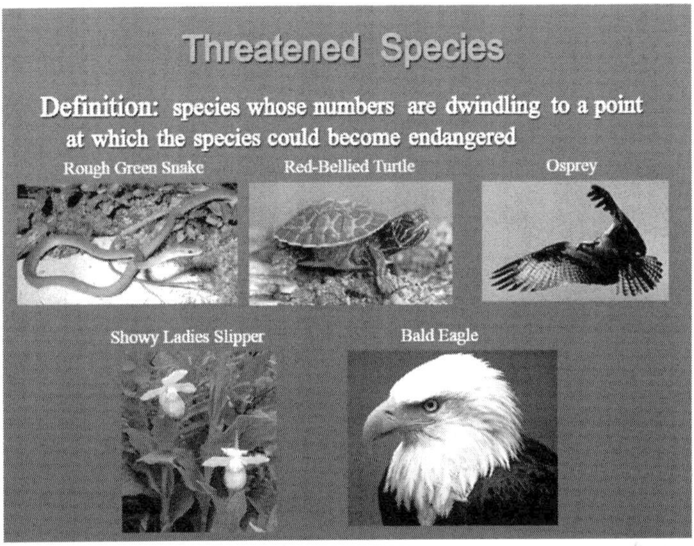

Figure 8.21: Threatened Species.

Source: https://slideplayer.com/slide/7400603/

Other reasons for loss of species are that wild animals are being hunted to extinction for bush meat, ivory, or "medicinal" products. The use of huge industrial fishing boats is damaging entire fish populations. Apart from their need to live, the services offered are vital as seen with bees.

Acid Rain

Acid rain is the by-product of air pollution because of the presence of pollutants in the climate. Corrosive downpour might be brought about because of the use of fossil fuels or volcanoes or spoiling vegetation which discharge Sulfur dioxide and Nitrogen oxides into the air.

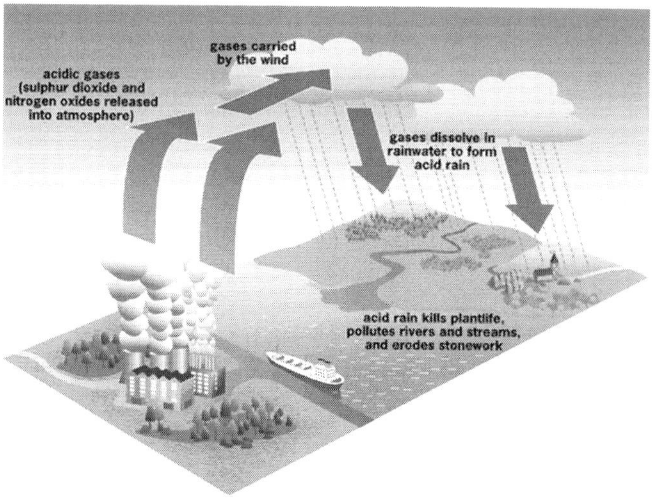

Figure 8.22: Acid Rain.

Source: http://inscrireair.tk/index/?1631501756857

Acid rain affects human health, wildlife and aquatic species. An increased acidity affects life of marine systems and trees by affecting their leaves.

Urban Sprawl

Urban Sprawl is a modern environmental problem that is rising as a challenge. Urban Sprawl means the uncontrolled expansion of urban areas. Urban sprawl is the relocation of the populace from high dense areas to low-density provincial zones to expand the city over more rustic area. People are moving to new locations and this is not a tough task anymore. Vacant lands are utilized for constructing building apartments and flats. Today it is very rare to see any vacant land.

Figure 8.23: Urban Sprawl in Northwest Calgary, Alberta, Canada.

Source: https://www.researchgate.net/figure/Urban-sprawl-in-northwest-Calgary-Alberta-Canada_fig4_313880885

Less vacant land means more environmental problems. The building of more houses leads to more pollution due to increased emission. An example is the face mask usage in China due to air pollution. Usage of concrete for buildings interrupts the natural water cycle, preventing rainwater from soaking into the ground. Urban sprawl brings about expanded movement, environmental concerns and well-being concerns. Example: The 2017 Hurricane Harvey in Houston.

Other impacts are degradation of land, traffic, and health issues.

Ocean Acidification

Ocean acidification is caused due to the increase in the Carbon dioxide level. In technical terms, Ocean acidification means continued lowering of the pH levels of the Earth's oceans due to the emission of carbon dioxide.

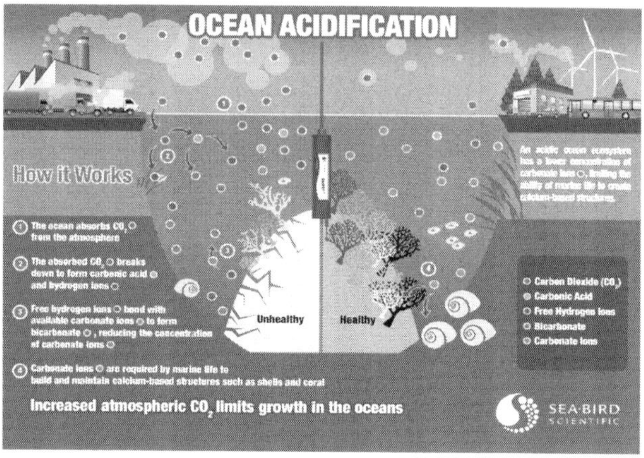

Figure 8.24: Ocean Acidification.

Source: http://www.deepmapscork.ie/environmental-priorities/climate-change/ocean-acidification/

Ocean acidification affects marine beings. The main impact of ocean acidification is on the shellfish and plankton in the same way as human osteoporosis. Marine food chain is disturbed and this may even lead to reductions in aquatic life. Destruction of coral reefs is one of the devastating effects of ocean acidification. Water becomes non-suitable for survival and the species start looking out for other habitats.

Overfishing

Figure 8.25: Overfishing.

Source: http://factsanddetails.com/world/cat53/sub340/item2196.html

Overfishing is also considered as an environmental issue and overfishing causes an imbalance in the ocean life and the natural ecosystem. The increased fishing causes a search for fish in new waters to keep the cycle going and affects coastal communities whose main bread and butter is fishing.

8.2 SUMMARY

Our environment is constantly changing and it will change more in future. No one can deny that. This change requires increased aware of the problems such as climate changes and temperature shifts.

Only when the problems are known, people can work on building up solutions for the same. For most of the environmental issues, humans are the main reason. Every study conducted on all major problems clearly proves this and this is undeniable. Protecting the environment is of utmost significance for the benefit of both the environment and humans individual, organizational or governmental levels.

There are many issues in our environment that needs to be solved. The role of many awareness campaigns across the globe is warranted to influence mindsets and conserve endangered species and habitats.

We are really in an emergency situation and we need to fix these issues from exploding. This requires rapid responses to avoid disaster. We have already discussed the various environmental issues that we are facing in the above section. Now let us see the ways by which we can either stop the problem from occurring or reduce the impact of them:

- We discussed more pollution. Pollution must be kept away and recovery is possible with care for ecosystems and sustainable resource usage. The easiest and fruitful way to get clean water to people is by digging wells. Water filters can also be used in places where there is less pollution. People must be educated on the causes and effects of water pollution so that everyone can work together to undo the damage humans have caused. Laws require stricter adherence across national borders.
- There are also ways to control water scarcity. Switching off taps when not in use and limiting the use of showers are also some beneficial ways to save water. When we understand that there is water scarcity, then we will automatically find ways to save water thereby eradicating water scarcity.

- In order to control global warming, the use of Carbon must be reduced to the maximum. Sharing transportation, switching off electrical items when not in use are some of the methods to reduce energy usage. The aspect of spreading awareness in this regard is highly warranted.
- Creation of waste materials can be reduced by using recyclable products or composting. Every household must take vow to reduce producing waste.
- Biodiversity loss can be tackled with awareness and making green choices.
- The remaining forests should be protected. Replacements of cut trees are vital.
- The use of less meat-based diets and sustainable options are warranted.
- Climate change can be reduced if we reduce the emission of greenhouse gases. Using renewable energy sources play an important role.
- The involvement of climate change in catastrophic issues is anticipated.
- The supporting of causes and legislation can aid in protecting the biodiversity species. An example is preferring organic products over paper and cardboard.

There is much more that can be followed in order to reverse the major environmental damage that requires strict adherence to avoid the Earth from getting uninhabitable in the future. There is nothing that is impossible as control is possible. By finding out correct paths and if we follow them sincerely, the environmental issues can be solved as humanity can have a direct and positive impact on the environment.

REFERENCES

1. https://www.conserve-energy-future.com/top-25-environmental-concerns.php
2. https://schooledbyscience.com/environmental-issues/
3. https://www.envirotech-online.com/news/air-monitoring/6/breaking-news/what-are-the-top-5-environmental-concerns-for-2019/47579

INDEX